MATERIAL MATTERS

NEW MATERIALS IN DESIGN

PHILIP HOWES AND ZOE LAUGHLIN

black dog publishing
london uk

200348
UCB

FOREWORD

Material Matters: New Materials in Design introduces new materials to designers. It is intended not just as a useful reference book for sourcing materials, but, with a broader scope, it aims to demonstrate how our material advances, built on our more commonplace materials, are shaping our designed world.

It would be impossible to make *Material Matters* an encyclopaedic compendium of all the materials available on the market today, instead Philip Howes and Zoe Laughlin introduce the reader to a level of materials science—an important part of understanding what materials are, how they come to be and why they behave as they do. The hope is that this book can be used to show not only what is currently available to the designer, but as inspiration, so that what will be possible with materials in the future might be imagined.

Each chapter—Metals, Glasses, Ceramics, Polymers, Composites and Futures—features an introduction by Howes and Laughlin, which is then followed by examples of materials, interspersed with interesting case studies that illustrate each material's use or development in contemporary design and engineering. The materials in each chapter are chosen to illustrate the variety available—each chapter begins with materials in their more familiar forms, exampled by their innovative uses, and become more complex or unusual. Many of the materials—both the commonplace and rare—are exampled by inspirational design projects. In doing so, *Material Matters* aims to give insights into the many ways, both conventional and more unusual, that designers work with materials, making everything from recycled clothing to expanding sofas, continually reinventing materials and in doing so revealing new and surprising properties.

OPENING PAGE
Detail of the patina created on newly antiqued glass. Image courtesy Antique Glass.

LEFT
Detail of a Velcro loop made from a polyester thread—an example of a material developed through biomimicry.

CONTENTS

60 GLASSES

***CASE STUDIES**

106 CERAMICS

130 POLYMERS

182 COMPOSITES

210 FUTURES

***CASE STUDIES**

INTRODUCTION

PHILIP HOWES AND ZOE LAUGHLIN

Materials matter. They are the stuff that constitutes everything around us; from mountains to mugs, everything is made from something. With such a broad defining remit, it is understandable that the question "what is a material?" is more commonly understood as "what are materials?", with generations of scientists attempting to categorise and understand the fundamental nature of matter to answer this question. Given their all-encompassing scope, it is not surprising that materials are also the subjects of much examination by non-scientists. Practitioners from arts, craft and design disciplines each have specific materials expertise and interests that lead them to experiment with materials and engage in a multitude of making activities. Be it a potter who works with a selection of clays and strives to perfect a type of glassy glaze, a musician who covets the specific tonal qualities of a particular wooden construction, or an artist who uses plastic detritus in great assemblages, materials are at the heart of many an artistic practice. The knowledge of how to manipulate materials to give specific forms or effects has historically been the property of the maker, passed down through generations and accumulating in cultures throughout globe. Be it methods to make beds, bread or buildings, the skills of masterful materials manipulation were practiced based. It was not until the invention of analytical technologies like microscopes that our understanding of what materials can be were informed by the internal reality of what materials are like. So successful were efforts to characterise and understand the behaviours of materials that the new found scientific knowledge of stuff now permeates much of our contemporary made world, therefore, for anyone engaged in a materials-based practice like design, a knowledge of the science of stuff is essential.

From a scientific perspective, the definition of a material starts at the smallest length scale, and works its way up from there. Although the wondrous scientific theories which describe the fundamental behaviour of matter are as intricate as they are complicated (quantum mechanics sets the stage here), it is possible for us to distil from them some simple and distinct concepts which underpin the character of all materials. We can start from the fact that all matter consists of atoms—as set out in the Periodic Table of Elements—consisting of nuclei containing protons and neutrons, orbited by electrons. The elements

THE PERIODIC TABLE

hydrogen 1 H 1.0079																		helium 2 He 4.0026
lithium 3 Li 6.941	beryllium 4 Be 9.0122												boron 5 B 10.811	carbon 6 C 12.011	nitrogen 7 N 14.007	oxygen 8 O 15.999	fluorine 9 F 18.998	neon 10 Ne 20.180
sodium 11 Na 22.990	magnesium 12 Mg 24.305												aluminium 13 Al 26.982	silicon 14 Si 28.086	phosphorus 15 P 30.974	sulfur 16 S 32.065	chlorine 17 Cl 35.453	argon 18 Ar 39.948
potassium 19 K 39.098	calcium 20 Ca 40.078		scandium 21 Sc 44.956	titanium 22 Ti 47.867	vanadium 23 V 50.942	chromium 24 Cr 51.996	manganese 25 Mn 54.938	iron 26 Fe 55.845	cobalt 27 Co 58.933	nickel 28 Ni 58.693	copper 29 Cu 63.546	zinc 30 Zn 65.39	gallium 31 Ga 69.723	germanium 32 Ge 72.61	arsenic 33 As 74.922	selenium 34 Se 78.96	bromine 35 Br 79.904	krypton 36 Kr 83.80
rubidium 37 Rb 85.468	strontium 38 Sr 87.62		yttrium 39 Y 88.906	zirconium 40 Zr 91.224	niobium 41 Nb 92.906	molybdenum 42 Mo 95.94	technetium 43 Tc [98]	ruthenium 44 Ru 101.07	rhodium 45 Rh 102.91	palladium 46 Pd 106.42	silver 47 Ag 107.87	cadmium 48 Cd 112.41	indium 49 In 114.82	tin 50 Sn 118.71	antimony 51 Sb 121.76	tellurium 52 Te 127.60	iodine 53 I 126.90	xenon 54 Xe 131.29
caesium 55 Cs 132.91	barium 56 Ba 137.33	57-70 *	lutetium 71 Lu 174.97	hafnium 72 Hf 178.49	tantalum 73 Ta 180.95	tungsten 74 W 183.84	rhenium 75 Re 186.21	osmium 76 Os 190.23	iridium 77 Ir 192.22	platinum 78 Pt 195.08	gold 79 Au 196.97	mercury 80 Hg 200.59	thallium 81 Tl 204.38	lead 82 Pb 207.2	bismuth 83 Bi 208.98	polonium 84 Po [209]	astatine 85 At [210]	radon 86 Rn [222]
francium 87 Fr [223]	radium 88 Ra [226]	89-102 * *	lawrencium 103 Lr [262]	rutherfordium 104 Rf [261]	dubnium 105 Db [262]	seaborgium 106 Sg [266]	bohrium 107 Bh [264]	hassium 108 Hs [269]	meitnerium 109 Mt [268]	ununnilium 110 Uun [271]	unununium 111 Uuu [272]	ununbium 112 Uub [277]		ununquadium 114 Uuq [289]				

* Lanthanide Series	lanthanum 57 La 138.91	cerium 58 Ce 140.12	praseodymium 59 Pr 140.91	neodymium 60 Nd 144.24	promethium 61 Pm [145]	samarium 62 Sm 150.36	europium 63 Eu 151.96	gadolinium 64 Gd 157.25	terbium 65 Tb 158.93	dysprosium 66 Dy 162.50	holmium 67 Ho 164.93	erbium 68 Er 167.26	thulium 69 Tm 168.93	ytterbium 70 Yb 173.04
** Actinide Series	actinium 89 Ac [227]	thorium 90 Th 232.04	protactinium 91 Pa 231.04	uranium 92 U 238.03	neptunium 93 Np [237]	plutonium 94 Pu [244]	americium 95 Am [243]	curium 96 Cm [247]	berkelium 97 Bk [247]	californium 98 Cf [251]	einsteinium 99 Es [252]	fermium 100 Fm [257]	mendelevium 101 Md [258]	nobelium 102 No [259]

differ by size (larger atoms have larger nuclei and more electrons), starting from the lightest element, hydrogen, all the way up to the heaviest natural element uranium. The characters of the elements fluctuate wildly as we ascend this list, ranging from stable and solitary elements like neon, to unstable and erratic ones like francium. In the natural world, however, very few of the elements exist in a pure state, which brings us to the second concept: the way atoms bind with one another. Atoms naturally arrange themselves into molecules, ranging from tiny to (relatively) massive, or into larger structures like crystals. The nature of such constructions is dependent on the strength and character of the binding between atoms, the way they are arranged, and the environment in which they find themselves. Finally, to obtain a solid material, these atomic constructions need to tightly bind together into a structure. If this binding is not very strong, the material will be a liquid, or if it is very weak, it will be a gas. An excellent example is water, a material with which we are familiar in all its states, from steam to ice. When the water molecules have lots of energy, as in steam, they move around barely even noticing each others' presence. However, as they lose energy and become slower, they begin to feel the attraction of their neighbouring molecules, forming liquid water. If they lose even more energy, the attraction between them becomes irresistible and they bind together to form ice. Overall, a point worth remembering is that the macroscopic properties of a material can always be traced back to the goings on at the microscopic level. Furthermore, a little understanding of what is going on down at this level will allow you to really understand and appreciate the wondrous nature of materials.

In order to characterise materials, it is necessary to define both composition (the types of atoms) and structure (how they are arranged). Materials scientists study material structure at different length scales and relate it to the material's overall characteristics. This multi-scale structure is present in all materials, including biomaterials with which we are maybe more familiar (DNA, in a cell nucleus, in a cell, in a tissue, in an organ, in a body etc.). The structure at all scales influences properties like elasticity, conductivity or strength, and it is a central tenet of materials science that a particular structure will always yield a particular set of properties. With a thought for materials engineering and design, it follows that the control of structure allows the control of material properties. This is of fundamental importance in materials: the better our understanding of, and control over, the microscopic world of materials, the greater is our ability to engineer materials in the macroscopic world.

After a particular material has been characterised by materials science, the data obtained becomes a source of information for those who need to choose a material with specific properties for a specific application. This is an extremely refined process, and it is now very well understood how specific physical properties relate to their overall behaviour. This means that there is a huge amount of technical materials information available scientists, technologists and industrialists, who use it to inform their decisions. However, this information is almost exclusively about the mechanical physicality of materials, properties like strength, ductility or hardness, and as such sheds very little light on an often understudied side of materials: their more sensorial aesthetic properties.

Within the world of materials practitioners, there exists a common division between the materials science community, those scientists, technologists and

Water—a material everyone is familiar with in its solid, liquid and gas states, comprised of a one oxygen atom bonded to two hydrogen atoms. It serves as a prime example of how the relationship between atomic and molecular building blocks defines the state of a material.

TOP RIGHT
As a gas—steam—the molecules are freely floating around, barely interacting at all. The molecules can feel each others' presence through electrostatic attraction between molecules as they lose energy and slow down, and the steam condenses into a liquid.

BOTTOM RIGHT
As a solid, it is possible to see with our bare eyes how the crystalline structure forms. The solid ice crystals 'grow' because of the network of bonds that form between the molecules. The same process occurs in all crystals, where atomic or molecular building blocks are bound together in a tight, ordered fashion.

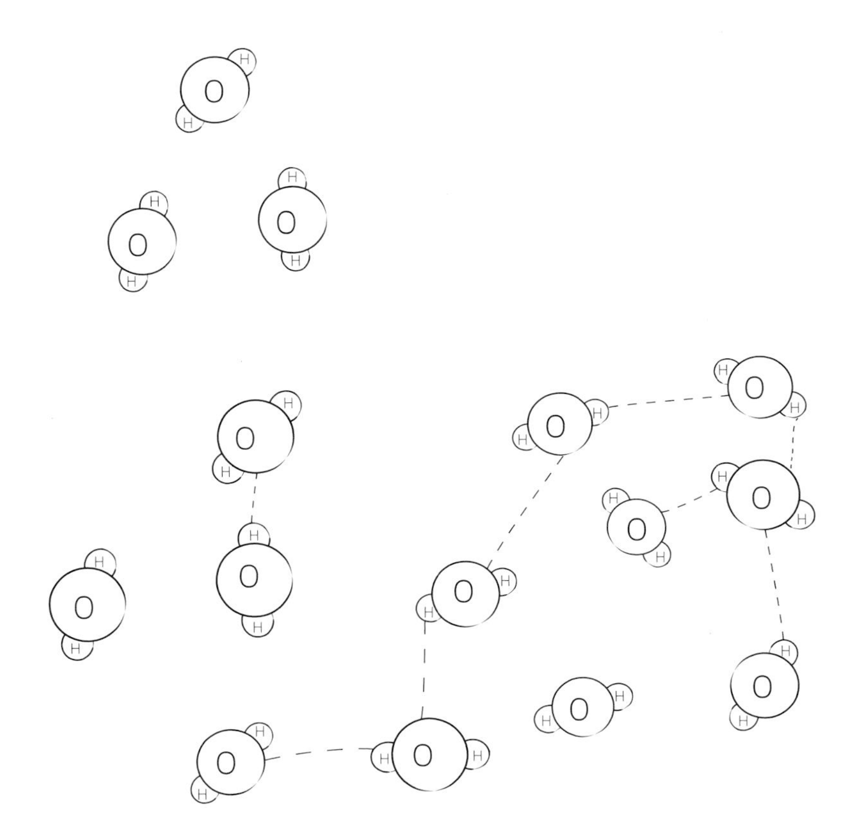

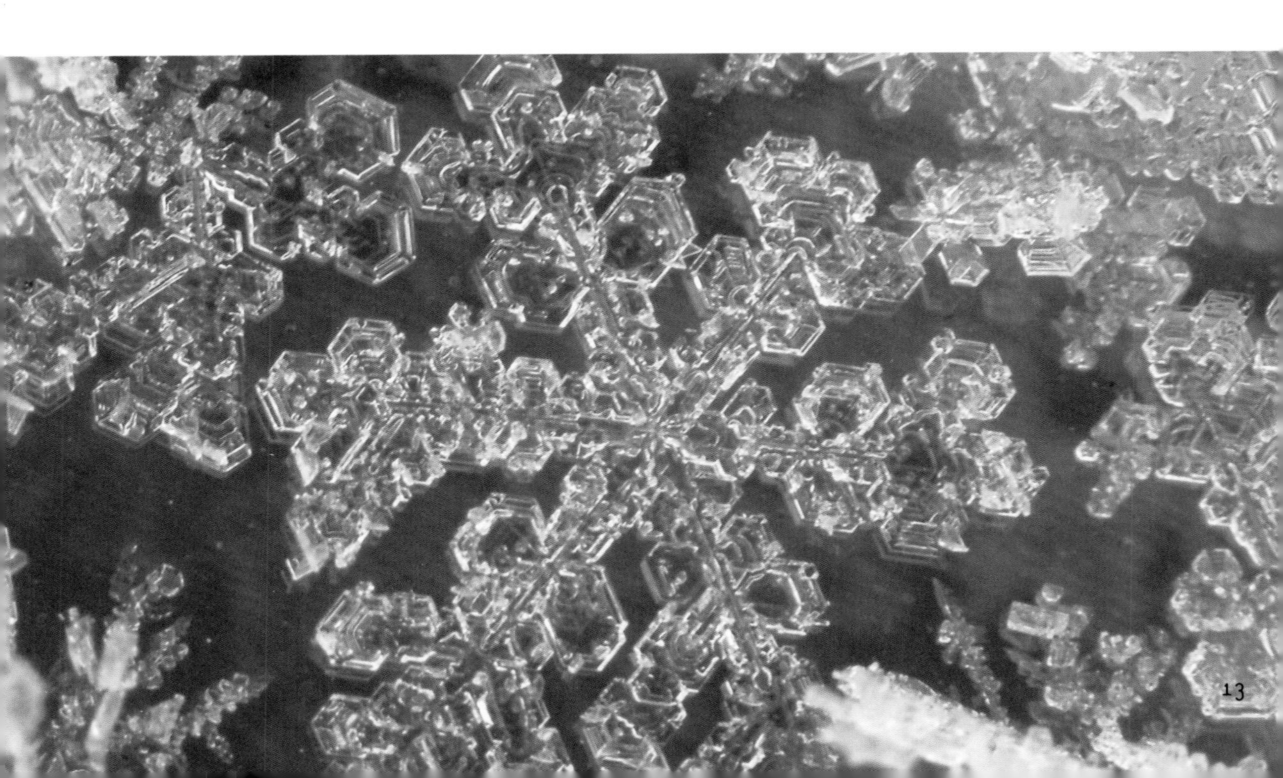

industrialists who are interested in the physicality of materials, and those in the materials-arts community who are interested in the sensorial and aesthetic properties, and our perception of them. Although both sides possess a huge amount of excellent materials knowledge, they often do not speak a common language. Some excellent designers have developed outstanding working relationships with engineers, but these collaborations are not embedded in the education of either party as part and parcel of what is required to fully understand and use materials in the world, through acts of making. However, we anticipate that people working at the interface of these two sides can alleviate this situation. By using scientific methods to study those properties of materials that are largely ignored by the materials science community yet are vitally important to the likes of industrial designers, architects, artisans and artists, it will be possible to develop a sensoaesthetic theory of materials that will act to bridge the science–arts materials divide. As the experience of the sensual and aesthetic side of materials is a matter of perception, a large part of this work falls within the realm of psychology. Hence, development here will require a strong fusion of materials science, a discipline driven by physical characterisation, and psychophysics, the science of perception.

Upon first consideration it may seem that a formal quantitative scientific discipline would not shed much light on the more experiential, qualitative and perceptual side of materials. However, upon closer inspection it becomes obvious that the way we interact with, and the resultant experience we gain from, all materials is rooted in their fundamental physical properties. Our advanced and highly sophisticated sense of materiality is something we take for granted. Each of us has a huge amount of materials knowledge that we use in all of our interactions with the physical world, and it is perpetually being updated. Our minds are like databases of sensory experiences, such that when we come into contact with a material the brain retrieves and pieces together the relevant information to interpret what is happening. Our sense of touch alone tells us a great deal of materials information. When touching a material, the dominant factors that we use to identify it are roughness, warmth and hardness. If we touch something that feels hard and cold, we expect it to be something like metal, glass or stone. If we add to this the surface texture, then we are more than likely going to be able to identify exactly what type of material it is. However, this example only considers touch in isolation and is therefore only the tip of the iceberg of how we use our sense to identify and form perceptions of materials in real world situations. When we use all of our sense in combination, we have some really powerful tools for materials investigation. Through our senses of touch, taste, smell, hearing and vision we can amass a huge amount of accurate information about a material very quickly.

So, where does materials science come into the equation? Well, all of our senses are actually detecting the physical properties of a material, and it is those physical properties which materials science sets out to investigate and characterise. For example, metals will generally feel cool to the touch because they conduct heat away from the skin very quickly. So we can say that, in general, materials with high thermal conductivity will feel cool to the touch. If an object is soft to the touch, then we can look at physical variables such as elastic modulus or plasticity to characterise the interaction. Thus it becomes apparent that our sensual and aesthetic interaction with, and appreciation of, materials is rooted in their physicality, and we can unravel these perceptual experiences using scientific methodologies.

We have considered what materials are from a scientific perspective, and we have thought about how our sensory appreciation of materials is dependent upon physical material properties. However, such an academic discussion of materiality really misses out a key point: materials are for making. The practical application of materials is what brings them alive: it is what turns plastic into prosthetic limbs, stone into sculpture, and metal into machinery. The materials in this book are presented within the traditional categories of science: Metals, Glasses, Ceramics, Polymers and Composites, plus a chapter of materials Futures. They are all, in their own right, incredible: stretching from the seemingly familiar to the sublimely sophisticated. However, going beyond these categorised materials into the design and making phase is what really turns them from standalone curiosities into revolutionary new ideas, objects and products. And it is in this step where the makers come to the fore, where tools are used to transform matter, and to create masterpieces.

METALS

Most people would know a metal when they come across it; hard, cold, shiny and perhaps grey in colour pretty much sums up our day-to-day experience of this category of material. So much so, that the adjective "metallic" is even used to describe things that are not made from metal at all but simply display a combination of metal-like characteristics. To be a metal however, is to be a material with some extraordinary properties that perform many of the key tasks in the world around us today. We can fly because of the science of aerodynamics, but we can fly fast using jet engines because of the science of metals.

The fundamental properties of metals result from the internal structures and behaviours of the material at differing scales of magnification. Starting down at the atomic scale, the classical way of thinking of all atoms is with electrons orbiting around a nucleus, rather like moons around a planet. When large numbers of metal atoms are clumped together, instead of the electrons of each atom orbiting their own nucleus, many are free to roam through the bulk of the material and create a sea of electrons that bind together the atomic nuclei. Such free electrons are the reason that metals are such good electrical conductors and are responsible for their reflective shine.

Metal alloys account for the vast majority of metals used in product design. They are manufactured by mixing molten metals together to produce new materials that exhibit altogether different properties to the original metals. A majority of the elements in the periodic table are metals, and this provides a wonderful palette for producing a vast number of different alloys, each with exquisitely tuned properties and characteristics. For example, if iron is

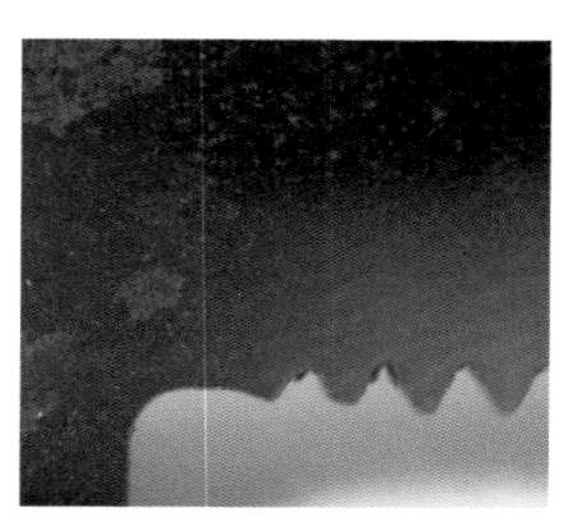
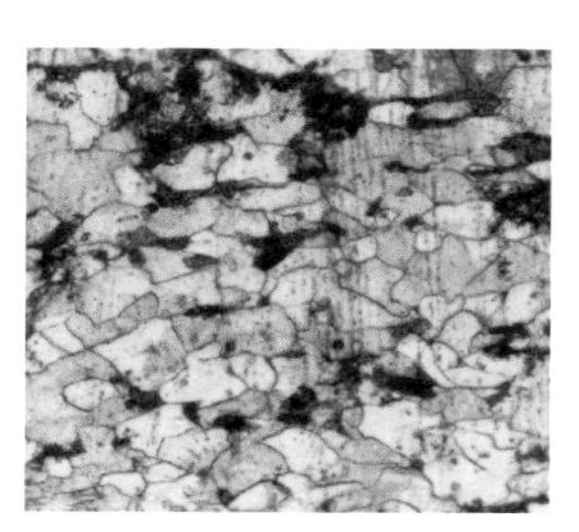
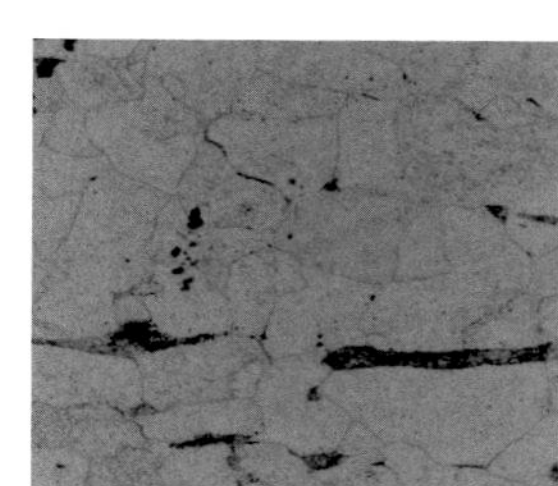

PREVIOUS PAGES
Detail of an aluminium extrusion accident, where molten aluminium was allowed to take its own form. As a solid it captures the metal's liquid state.

ABOVE
A bolt sawn in half reveals its metallic crystalline grain structure when gradually magnified and viewed using an optical microscope.

mixed with chromium to around ten per cent and a sprinkling of carbon, it yields stainless steel, an alloy with excellent corrosion resistant properties not present in pure iron.

Moving up the length scale (zooming out from the atomic level), a further remarkable aspect of metals is revealed; that is, that all metals are made from crystals. To many unfamiliar with the internal workings of metals, this is somewhat surprising and counterintuitive as crystals are most commonly associated with more transparent substances like ice, but the process of the solidification of metals is one of crystallisation. In a hot molten metal, the atoms have no fixed position, having only fleeting relationships with each other as they move around in the sea of electrons that they share. As the liquid metal cools, some of the atoms begin to bind together, gradually growing into many tiny crystals that spread throughout the liquid. As these crystals grow they consume all of the liquid, causing the metal to solidify as a fused mass of individual crystalline grains. The characteristics of individual metals are largely defined by the size and shape of these crystal grains, and on how the atoms bind and behave within them.

Functionally, metals are often used for their strength or conductivity. However, they have a definite senso-aesthetic quality that affords function far beyond any mechanical or scientific use. For example, architects can make modern buildings where the sweeping curves and shining surfaces of metal frameworks and facades provide equal amounts of structural integrity and visual interest. The specific senso-aesthetic quality of metals can be manipulated through processes and treatments to produce both visual and haptic effects that provide a range of cultural references and connotations.

PH and ZL

The sweeping curves such as in iconic buildings like Frank Gehry's Walt Disney Convention Center are made possible by the science and properties of metals.

DUCTILE IRON

PROPERTIES

More elastic than cast iron

Strong

Tough

Can be engineered for different properties

APPLICATIONS

Pressured pipes

Engine parts

Manhole covers

Sculpture

INFO

www.saint-gobain-pam.co.uk

www.ftductile.co.uk

www.castings.plc.uk

www.ductilecastiron.com

Ductile Iron was discovered in the 1940s, and since then its improvement has led to its use in a wide variety of applications, from engine parts to drainage. It is similar to cast iron except with several materials added at its liquid stage, making it more flexible, less likely to corrode and less brittle.

Carbon, silicon, manganese, magnesium, phosphorus and sulphur are usually added in small amounts, with slight variations in composition creating a variety of ductile iron materials with different properties.

A recent torque test has shown how Ductile Iron rails can be bent and twisted 90 degrees before breaking, demonstrating how much more elastic it is than regular cast iron.

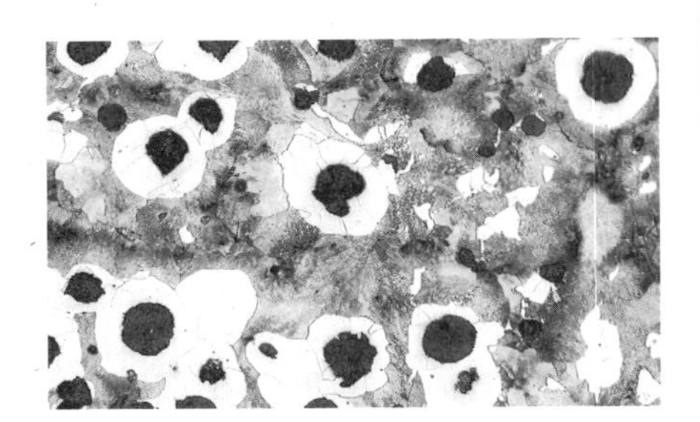

TOP LEFT
Polished and magnified Ductile Iron shows the presence of graphite particles in the metal matrix.

TOP RIGHT
Billets are created using the same methods as conventional cast iron, creating a variety of different parts for machinery, street furniture and systems. Image courtesy Ductile Cast Iron.

BOTTOM LEFT
A selection of parts produced in Ductile Iron by the Durham Foundry. Image courtesy Ductile Cast Iron.

BOTTOM RIGHT
Tensile tests demonstrate the material's elasticity before breaking.

ADDITIVE LAYER MANUFACTURING

Additive Layer Manufacturing (ALM) uses 3-D printing to build structural forms. It works by sequentially adding very thin incremental layers of material on top of each other and bonding them with the heat of a laser. Numerous metals can be used, such as titanium, aluminium or stainless steel, and the resulting structures can replace traditionally cast or superformed shapes.

This process is capable of creating highly complex forms and shapes, much more intricate than can easily be achieved from traditional casting processes. There is also very little waste as the parts are formed from the 'bottom up', additively, so there are no off-cuts. This is in stark contrast to some traditional machining processes which can waste as much as 90 per cent of their raw materials.

In working with its parent company EADS to develop parts for the A380 jet airliner, Airbus has identified over 1,000 parts that could be manufactured using ALM. This has the potential to considerably reduce the airliner's weight as parts can be made with only the structurally integral elements in place, thus saving huge amounts of fuel. A 200-watt laser is used to sinter the materials, which takes up to two hours to produce usable parts, but does not require huge amounts of power as compared to melting and moulding metals, therefore also saving time and energy in the manufacturing process.

PROPERTIES

Efficient

Rapid production

Weight reduction technology

Manufactures difficult geometry

APPLICATIONS

Aerodynamics

Structural forms

Replaces cast or superformed shapes

INFO

www.eads.com

www.impc.org.uk/alm.php

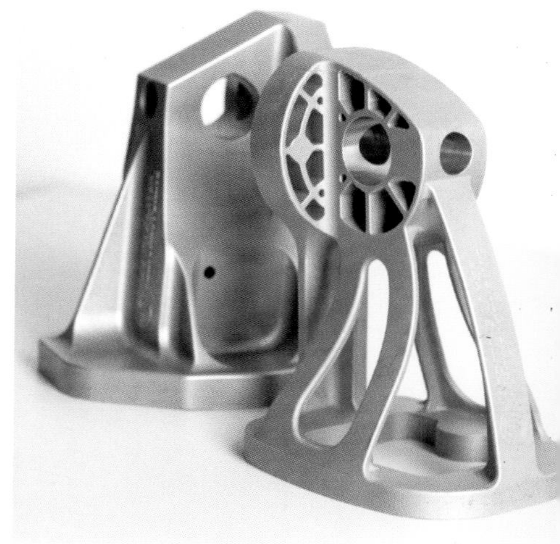

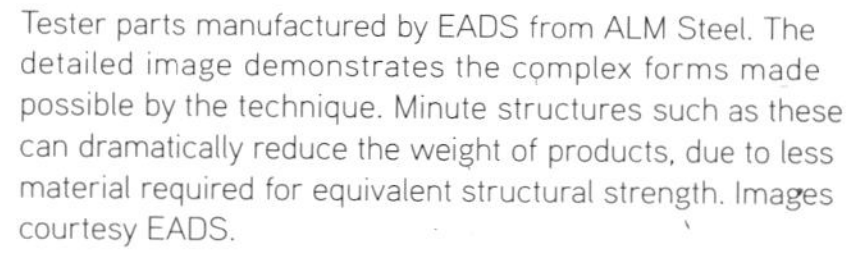

Tester parts manufactured by EADS from ALM Steel. The detailed image demonstrates the complex forms made possible by the technique. Minute structures such as these can dramatically reduce the weight of products, due to less material required for equivalent structural strength. Images courtesy EADS.

CASE STUDY

SHAPING WITH MAGNETIC FORCE

THE GRAVITY STOOL

Designer Jólan van der Wiel created this tripod stool from a mixture of iron filings and plastic. Like many successful projects using inherent material properties to good effect, the stool straddles the genres of art, design and manufacture. Van der Weil says, "I admire objects that show an experimental discovery, translated to a functional design."

The stools are made by mixing together iron filings with plastic in its liquid phase and using magnetic force to shape the mixture as the plastic solidifies. The mixture is then placed underneath three large magnets, which are raised up by van der Wiel and, when timed just right, cause the iron filings in the mixture to rise up and form the three legs of the stool. The stools can hold around 200 kilograms or 441 pounds and come in yellow, purple, grey and back, as well as many different shapes and heights.

Van der Wiel constructed the machine from scratch, a contraption using basic materials and requiring no electricity. This impressive invention, showcasing ecological awareness, earned him first prize at the D3 Design Competition in the Interior Innovation Award 2012 at Imm Cologne's contest for young talent, hosted by the German Design Council. Their motto, "For the very best to become standard tomorrow, you have to select something special today", proves the success of the overall concept and longevity of product design.

LEFT
A finished Gravity Stool.

OPPOSITE TOP
A detail of the grainy texture achieved from the combination of iron filings and smooth plastic.

OPPOSITE BOTTOM
The process of raising the magnets drags the form upwards using magnetic force. Van der Wiel manually raises the powerful magnets, making each stool unique. Images courtesy Jólan van der Wiel.

BRASS

PROPERTIES
Antimicrobial
Aural amplitude
Generates little friction

APPLICATIONS
Ammunition casing
Bearings
Clock-making
Door handles
Gears
Musical instruments

INFO
www.brass.org
www.gears-manufacturers.com
brass-gears.html
horologistoflondon.com
store.brassmanbrass.com

TOP
Brass gears are still used in contemporary clock-making and in gears for ammunition. These clock gears are reproduced here double their actual size.

BOTTOM
A brass sample showing the typical colour of the material.

Brass is an alloy of copper and zinc, and is formulated in many variations to produce a gold-coloured metal that can be machined to relatively fine tolerances. The fact that Brass can be engineered with such precision has led to it being used for centuries by horologists in the making of clocks. It also generates very little friction between moving parts made of the material. It is therefore still the ideal material for gears, bearings and the casing of ammunition, where any excess heat or rogue spark could be disastrous.

Brass was traditionally a material used to create door handles, due not only to the ease with which it can be machined but also the aesthetic effect it produces, especially when polished. Over time fashions change, new materials enter the market and shiny brass door handles and push-plates are now less common. However, recent research has found that polished Brass exhibits extraordinary antimicrobial qualities, suggesting that we may see a resurgence in the use of brass door handles, especially in public buildings and hospitals.

An iconic use of Brass is seen in musical instruments, where an entire section of the orchestra takes its name from the material. The Brass tubing of a trombone, for example, is bent into shape by filling a length of strait brass tube with water, freezing it to ice and then forming the desired curves by bending it round an iron form of the desired dimensions. The fact that the tube is filled with a solid prevents it from collapsing when bent, insuring that the internal volume of the instrument is maintained evenly throughout. Although physics would tell us that the sound produced by a trombone is simply due to the vibrations of a column of air inside the instrument, a trombonist will 'speak up' for the material and tell you that playing a plastic trombone produces a sound of strikingly different qualities to a Brass trombone. This is due to the fact that the brass tubes that encase the columns of vibrating air, in term vibrate as air passes through them. This vibration of the Brass produces a specific yet extremely subtle resonance, pitch and sustained note, meaning that it is not only the object but also the material that is musical.

ECO BRASS

Eco Brass is a lead-free brass alloy that is reportedly more environmentally friendly than those brass alloys that contain lead. Developed by Mitsubishi Shindoh Co., Eco Brass is unique in that it does not suffer from the usual problems that face most brass materials, such as stress corrosion cracking and dezincification corrosion.

Eco Brass contains silicon instead of lead, which not only improves its health and enviromental credentials, but also provides it with excellent machinablility. Other advantageous properties of Eco Brass include its forgeability and high strength, equivalent to that of stainless steel.

PROPERTIES

Environmentally friendly

High strength

Good machinability

Excellent hot forgeability

Good castability

Excellent de-zincification corrosion resistance

Excellent stress corrosion cracking resistance

Excellent warm brittleness cracking resistance

APPLICATIONS

Automobile parts

Electronics

Water supply devices

INFO

www.ecobrass.com

www.mitsubishi-shindoh.com

Samples of Eco Brass which have been machined, demonstrating the material's excellent machinability in comparison to regualr brass.

ANTIMICROBIAL COPPER

PROPERTIES

Antimicrobial

APPLICATIONS

Prevents the spread of infection in hospitals

INFO

www.antimicrobialcopper.com

www.architecturalbronze.com

Antimicrobial Copper is reportedly the most effective touch surface material in the fight against pathogenic microbes. It is able to kill more than 99.9 per cent of bacteria within two hours of their exposure to it by rupturing the cell membranes through metal ion interactions.

A potentially important use for this material is in hospitals, as it would aid in combating the spread of infections such as influenza, MRSA and salmonella enteritidis, among others. This is backed up by extensive testing which has shown that if surfaces such as door handles and table tops are made from antimicrobial copper, a serious reduction in hospital infections would be possible.

Copper plated handles and soap dispenser lids from Colonial Bronze. By replacing standard hospital fittings with these products, the spread of hospital viruses can be drastically reduced. Image courtesy Colonial Bronze.

COPPER TAPE

PROPERTIES

Inexpensive

Attractive solution

Harmless to the environment

INFO

www.convertape.com

Slugs dislike copper. Upon contact with it they experience an electrical shock due to a reaction between their slimy undercarriage and the copper surface. Convertape produce a layer of ultra thin tape, which can be wrapped around plant pots to repel slugs.

Copper tape looks just like a roll of sticky tape and is as easy to apply as it is incredibly thin and has an adhesive backing. Image courtesy Convertape.

SOFT CIRCUITS

Engineers at the University of Illinois have developed a device platform that combines electronic components for sensing with a skin-like patch that is very slim and easy to attach. In fact, they can even be concealed under a temporary tattoo, covertly collecting data or transmitting signals.

The fine circuits are woven, allowing them to bend and twist whilst they function. The researchers tested their technology with sensors, LEDs, transistors, wireless antennas and solar cells, amongst other devices, demonstrating its array of potential functions, from military to home computing.

This technology is already being commercialised. As the group adapted techniques from the semiconductor industry the processes required are easily scalable for bulk manufacture.

The devices could be used to gather data by taking measurements from the body and feeding data to a computer. Sensors may be used to monitor speech or muscle movement, allowing the potential for controlling virtual reality devices in real time through movement alone.

PROPERTIES
Discreet
Easily attachable
Thin
Flexible

APPLICATIONS
Data collection
Home/Military computing
Potential for bulk manufacture
Signal transmission

INFO
www.news.illinois.edu

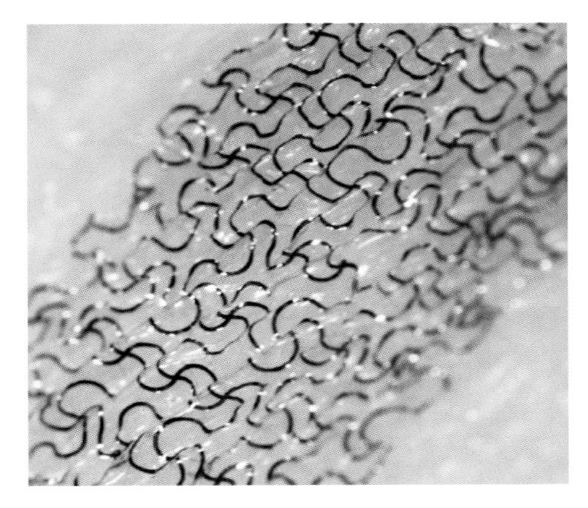

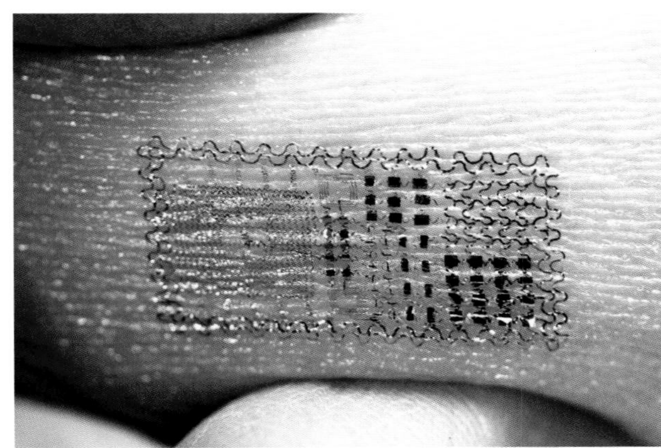

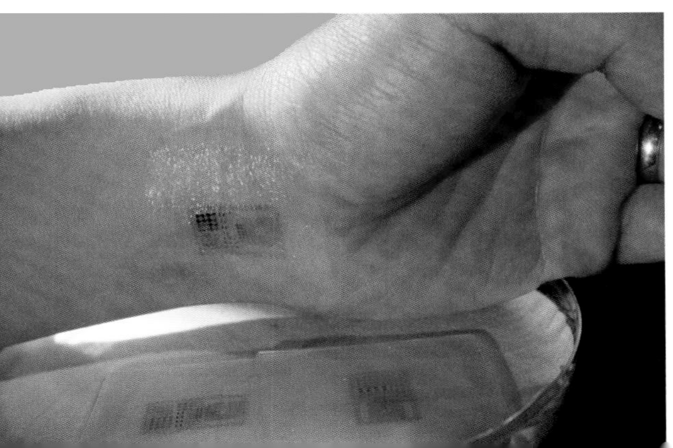

TOP LEFT
Detail of a soft circuit.

BOTTOM LEFT
An ultrathin, electronic patch with the properties of skin, applied to the wrist for EMG and other measurements.

RIGHT
Electronics mounted directly onto the skin, with no need for wire or conductive gel. They bend, stretch and deform with the same mechanical properties of skin, granting the wearer comfort and freedom of movement. Images © John A Rogers University of Illinois.

RECYCLED COPPER

PROPERTIES

Resistant to corrosion

Excellent thermal and electronic conductivity

APPLICATIONS

Construction

Electrics

Plumbing

INFO

www.copper.org

www.copperinfo.co.uk

Copper is an important material in the building, electrical and plumbing industries. Crucially, it can be recycled relatively easily, with its quality remaining constant and not deteriorating with repeated processing, unlike many plastics.

Some of Copper's many advantages include its high weather-resistance, its distinctive appearance and its anti-bacterial properties. This makes it a popular material for a number of different uses, from coins to architectural cladding.

Around 80 per cent of Copper mined since antiquity is still in use. Over 40 per cent of recycled products use Copper.

Much of our plumbing pipes and wiring are still made from copper. Because of the value of copper, this is almost all recycled and in turn our used copper pipes and wires are made into new products.

OPPOSITE
The Irving Convention Center, designed by RMJM Hillier's New York Studio, is one of the most environmentally friendly convention centres in the world. It is 'wrapped' in recycled copper plate.

ALUMINIUM FOILS

PROPERTIES

Reflective

Relatively low melting point

Shiny and matte

Ability to compress into very thin sheets

APPLICATIONS

Culinary

Duct sealing

Electromagnetic shielding

Thermal insulation

INFO

www.nfpfoil.com

www.scapa.com

Aluminium can be drawn into extremely thin sheets (standard household foil is only 0.016 millimetres thick), and is made by continually casting and cold rolling the metal. Aluminium has a lower melting point in comparison with harder metals like iron, and therefore the pressure of cold rolling is enough to induce significant softening, allowing subsequent compression.

Aluminium Foil has one side which is shiny, while the other is matte. This occurs because the foil is doubled-layered on its last trip through the rollers to achieve double compression. The sides exposed to the rollers become shiny due to annealing, whilst the sides in contact with each other remain matte.

Besides being useful for culinary purposes, Aluminium Foil is often used for thermal insulation and electromagnetic shielding (due to its reflective properties), and in industrial applications it is often used to polish steel. Aluminium foil dipped in water removes rust from steel through a reaction initiated by heat generated through friction. The very thin sheet of Aluminium gets warm and oxidises to produce aluminium oxide, which has higher reduction potential than iron, therefore leaching oxygen atoms away from the rust on the steel's surface.

BLACKWRAP

PROPERTIES

Light-weight

Malleable

Matte black surface

Temperature resistant

APPLICATIONS

Masks light

Television and theatre production

INFO

www.filmtools.com

www.gamonline.com

www.videogear.co.uk

Blackwrap is a high temperature resisting light masking Aluminium Foil. Its matte black surface reflects very little light and inhibits the passage of all light through the material. The fact that it is a thin yet sturdy foil means that it can be shaped easily and holds the form it is moulded into.

Used in the theatre, film, and television industries for controlling light propagation, direction and shape, it can be placed in close contact with extremely hot equipment without melting or deforming. For example, holes can be cut into the foil to create masks of specific shapes that can be place in front of lights to project shafts of light of specific dimensions onto the surface of a stage. The flexibility of the material enables it to be repeatedly and easily moulded, making it useful in situations where improvised lighting solutions are required.

TECHNICAL ALUMINIUM FOIL

PROPERTIES

Incombustible

Reflective

Corrosion resistant

APPLICATIONS

Architecture

Fibre optic and electrical cables

INFO

www.hydro.com

Aluminium Foils are used as fire resistant barriers for insulation because they are incombustible and their shiny surfaces reflect heat. Additionally, they are resistant to corrosion, making them ideal for insulation layers in roofs or in between walls, where gaining access to mend eroded areas once the walls are covered would be difficult. Aluminium Foil is an important layer in fireproof doors as a part of fire regulation codes because it also prevents oxygen from passing through.

In industry, Aluminium Foils are used in electrical cables as both an insulator against moisture and corrosion, as well as magnetic and radio frequency emissions.

LEFT
Rolls of Aluminium for use as fire doors or in Superforming. Image (c) Norsk Hydro.

ALUMINIUM TAPE

3M™ make a slim and soft Aluminium Tape called "Aluminium 425", which is usually used for wrapping pipes, repairing damage and sealing sensitive surfaces from dust and moisture. The tape is highly reflective and shiny and it is easy to imagine it having more aesthetic or sculptural applications.

As it is made from extremely thin aluminium it is thermally conductive and resistant to weathering and UV rays. It is therefore good for both indoor and outdoor applications.

PROPERTIES

Thermally conductive

Moisture and solvent resistant

Heat and light reflective

Flame resistant

Outdoor weathering and UV resistant

APPLICATIONS

Wrapping for pipes

Protective layering

INFO

www.Shop3M.com

SUPERFORM ALUMINIUM

PROPERTIES

Can be tooled like regular aluminium sheet

Enables complex geometric 3-D forms in metal

APPLICATIONS

Architecture

Automotive

Aerospace

INFO

www.superform-aluminium.com

www.luxfer.com

www.rogiersterk.nl

Superforming, also known as "superplastic processing", is a heat moulding process. Although it can be used to process a number of materials, it has become particularly associated with Aluminium and is used extensively in the automobile industry.

Aluminium sheet is heated to a temperature between 450 and 500 degrees Celsius or 842 and 932 degrees Fahrenheit, at which point it becomes pliable. Aluminium structures can then be formed by controlling the air pressure either side of the sheet, for example to vacuum mould around a preformed tool, or blowing the sheet into a mould. As several different techniques can be used, it is possible to create intricate three-dimensional shapes from a single sheet of Aluminium. The process also allows designers more freedom to create varying geometric forms and smaller batches of complex designs for furniture and architectural panels.

Superform, a company specialising in the process, also use it for airframe alloys and carbon composites and are thus able to create very high specification products for the aerospace industry.

Two views of Rogier Sterk's Tai Chi light, above, from the side and right, from below. The superform process makes the geometric pattern easy to form from sheet metal.

High-end cars produced in small quantities use superforming to shape the aluminium bodywork. Here the Morgan Aeromax parts have been formed using superforming and are awaiting painting. Photographs by exfordy.

ALUMINIUM HONEYCOMB

There are many different forms of Aluminium Honeycomb, from regular hexagons to reinforced or elongated shapes, as well as many variations in the type of aluminium used. Each variation opens up another application for this versatile material.

Strong and light, it is often used as the central component in composite sandwich panels for architectural applications. The combination of the strong composite panel sides adhered to the light structure makes for a very solid form with both insulating qualities and structural integrity. Without reinforcement the material is used as a shock absorber.

In general, the tighter the arrangement of hexagons the more high-tech the application. Aluminium Honeycombs used for aerospace applications are under half a centimetre in diameter whereas the size of the gaps in a panel in a domestic environment are generally around two centimetres, making the domestic grade cheaper as it requires less material, but the aerospace variety much stronger.

PROPERTIES

Completely recyclable
Good compression strength
Good shear strength
Good vibration dampening

APPLICATIONS

Architecture
Insulation
Shock absorbers
Aerospace

INFO

www.easycomposites.co.uk
www.euro-emc.co.uk

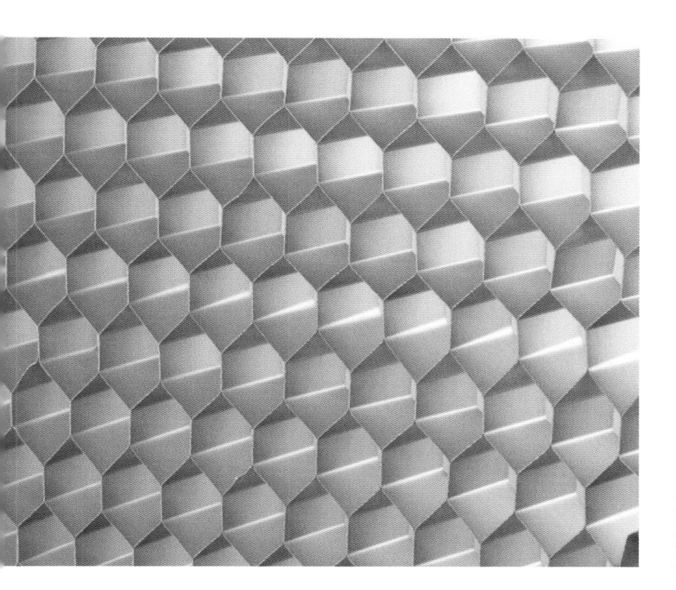

OPPOSITE
The tight arrangement of hexagons in a very flat shape allows the usually very stiff material to flex.

ABOVE
Aluminium Honeycomb sample. The larger, even-sided structure is mainly used in architectural applications.

RIGHT
Increasing the depth of the honeycomb creates a stiffer material, capable of great structural strength along its length but if compressed, the honeycomb shapes will deform easily because of their thin walls.

CASE STUDY

CUTTING AWAY

MILLING METAL FORMS

Drummond Masterton uses four-axis CNC (Computer Numerical Control) milling. Challenging on a technical basis, the "production requires the development of new methods for creating user-defined CNC toolpaths to machine the design from one solid billet of aluminium". The process involves shaping a solid block of Aluminium to a shape generated in 3-D software from 2-D terrain data, removing sequential layers of material to develop the textures of his forms.

For his *Terrain Cup*, Masterton tested a range of different scales for maps through Digimap Carto to get the right level of detail. It was established early on in the research project that the scale of 1:25,000 provided detail that was unrealistic from a computing power perspective. 1:50,000 provided enough contour detail to resemble the landscape. The particular surface finish of *Terrain Cup* was achieved through making a number of fine cuts. The test pieces were refined by printing them in 3-D, using a Z Corp Rapid Prototyping machine to check scaling and alignment of the adjoined parts.

Masterton trained as a 3-D designer and is now a key player on the British contemporary craft scene. He has exhibited in the UK and Europe for the past decade and is a Senior Lecturer in Three-Dimensional Design within the 'Autonomatic Research Cluster' at University College Falmouth.

OPPOSITE
Decagon Bowl milled in aluminium using CNC milling.

TOP AND BOTTOM
Textural test samples to determine the amount and depth of milling required. Such details demonstrate the depth precision of the method. Each successive depth can be precisely controlled, varying from a form with smooth transitions to quite sharply stepped increments.

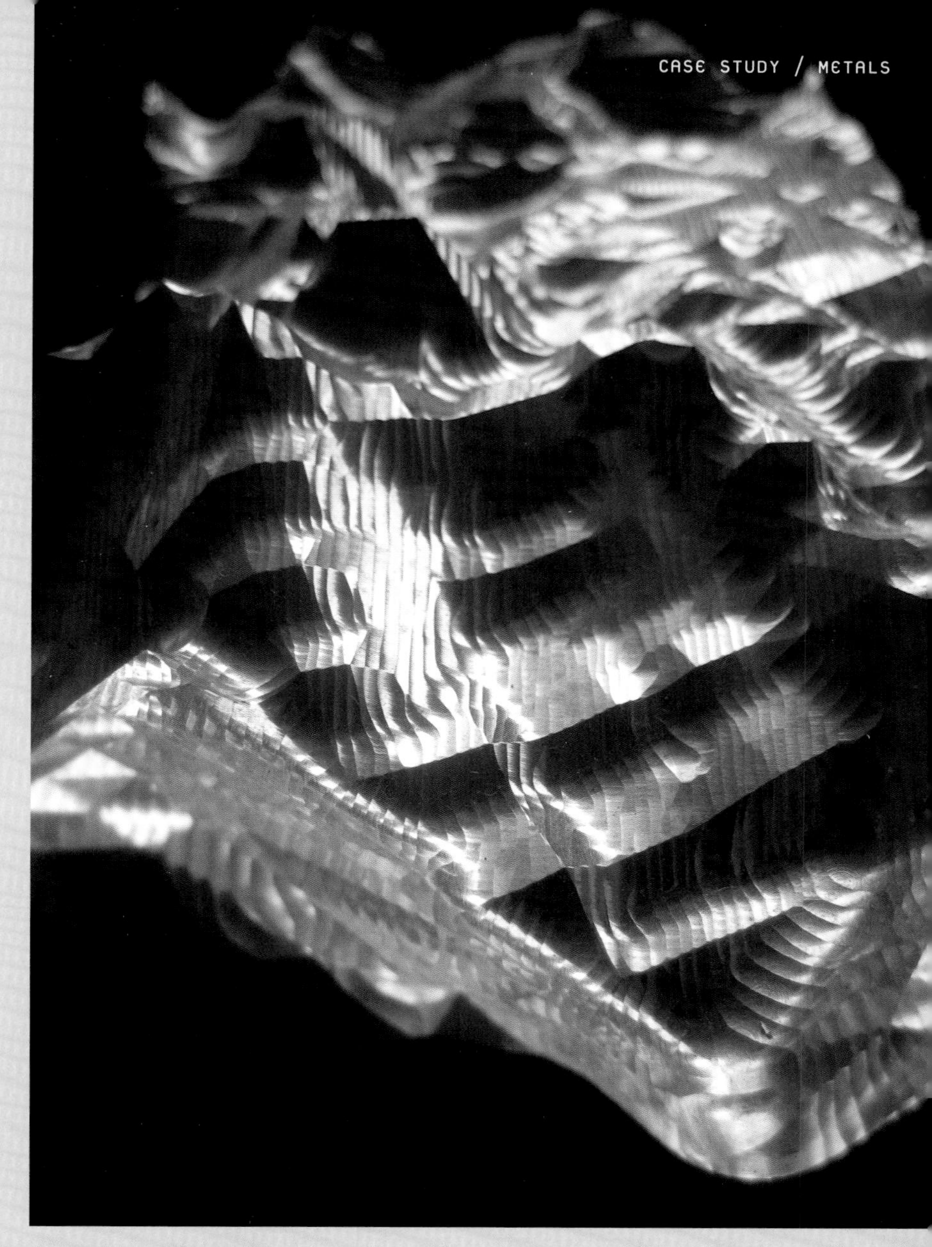

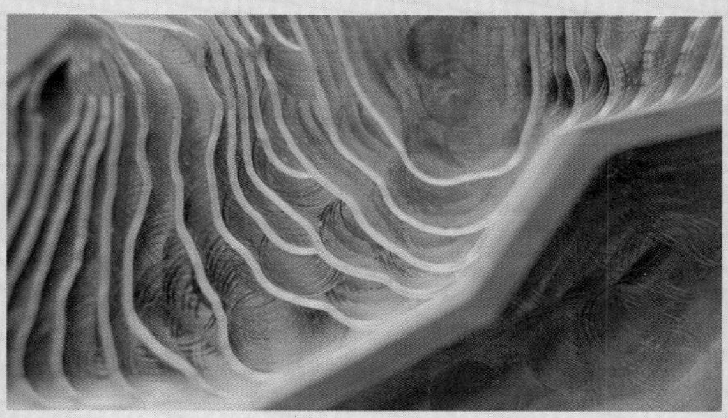

TIN

PROPERTIES

Malleable

Resistant to corrosion

Superconductive

Zero electrical resistance

Lightly aligned crystal microstructure

APPLICATIONS

Architecture

Design

Canning

INFO

www.chemicool.com

www.tinwaredirect.com

Mainly obtained from the naturally occurring mineral casserite, Tin is a malleable and silvery metal that is often used in alloys such as pewter and bronze. It does not easily oxidise in air and is corrosion resistant, making it an important material for the canning industry where liquids of different acidity levels are required to sit in contact with a metal for a prolonged period of time. Despite the name, "tin cans" are in fact steel cans coated with a thin layer of Tin to prevent the contents reacting with the container.

If you bend Tin sticks such as those, illustrated below, they make a noise described as the "cry" of tin. If you were to look at Tin under a microscope you would see why. Its microstructure is similar to that of diamond, silicon and germanium, with tightly aligned crystals that scrape across each other when the stick is bent, producing vibrations that you can not only hear, but also feel.

Tin is a very commonly used element in architecture and design. For example, it is the main constituent of most solders, which are crucial in joining steel in skyscrapers, and also in securing components in electronic devices. It is also used to make large panes of glass during the float glass process where molten glass is floated on a bed of molten Tin, allowing a thin and uniform layer of glass to form and solidify, maintaining it as an extremely important material in design today.

Below a temperature of 3.7 Kelvin (almost -270 degrees Celsius or -454 degrees Farenheit) Tin exhibits a fascinating property known as superconductivity. Apart from exhibiting zero electrical resistance, superconductors can be used to levitate magnets through a process called the "Meissner Effect." Tin was one of the first materials found to exhibit this behaviour, as discovered by Robert Ochsenfeld in 1933.

BELOW
Tin sticks, which when bent exhibit a "tin cry" sound.

OPPOSITE
A Stainless Steel Thread coated in silk. The fine thread allows the silk to hold its shape, and even makes it magnetic. Image courtesy Habu Textiles.

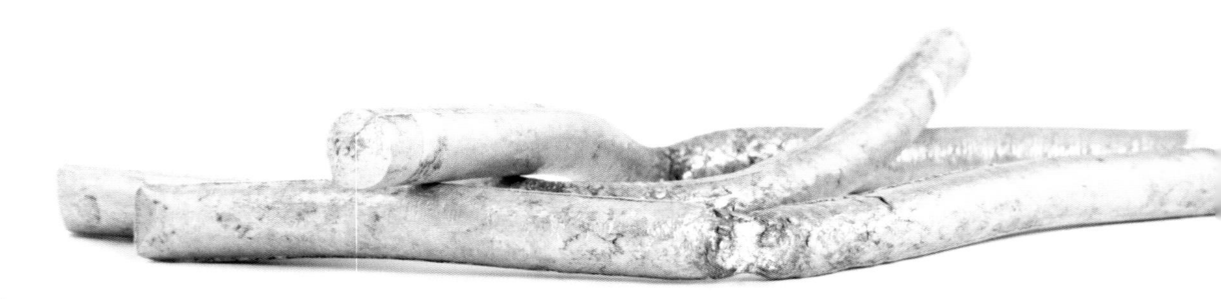

STAINLESS STEEL THREAD

Habu Textiles in Japan produce Stainless Steel Threads wrapped in silk. The stainless steel gives the yarn a degree of rigidity, allowing it to be manipulated and hold its shape.

Although it is currently mostly used in knitting to create decorative silk garments with dramatic shape-holding properties, given its strength and fine diameter (it is only 40 microns wide), it has many other potential uses.

With a core of steel, the fine thread also responds dynamically when in contact with a magnet.

PROPERTIES

Fine diameter
Magnetic
Shape-holding
Tough

APPLICATIONS

Knitting

INFO

www.habutextiles.com

CONDUCTIVE TAPE

PROPERTIES

Adhesive

Elastic

Weather resistant

Electrical resistance per metre measures 0.15 to 0.03 OHM, depending on the tape used

APPLICATIONS

Woven or warp-knitted narrow fabrics

INFO

www.amohr.com

Made from elastic yarn, A-Mohr's Conductive Tapes contain tin plated copper strands. The strands hover in wavy lines along the tape so that when stretched, the elasticity of the material deforms the tape without damaging the wires. The larger of the tapes is backed with a 'hotmelt' material, which allows it to be ironed onto almost any surface.

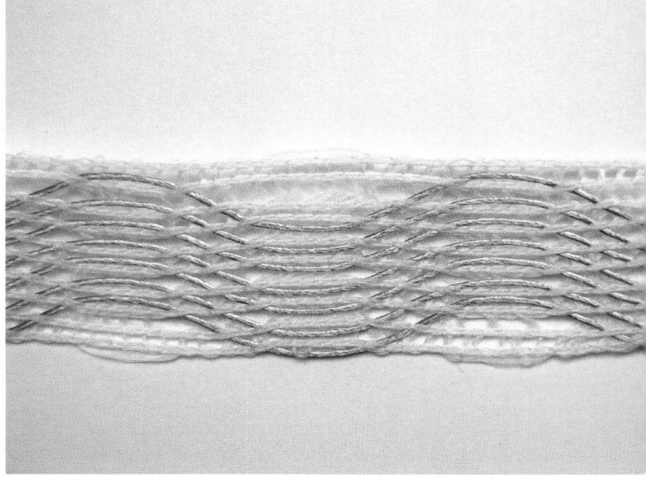

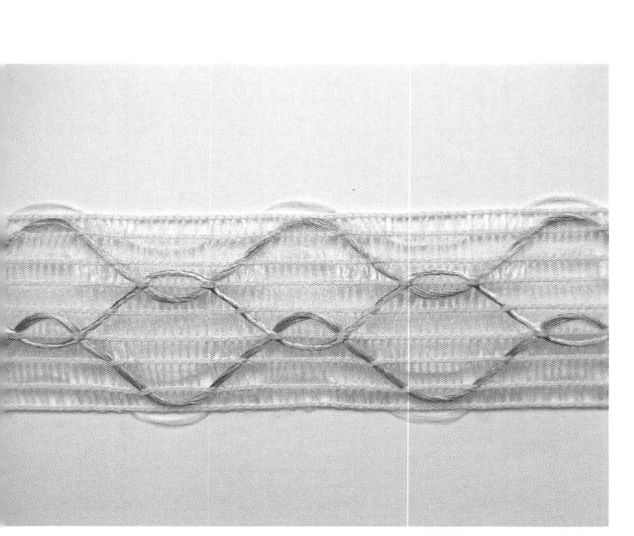

Conductive Tapes in various formations. The tightness of the waves affects the material's elasticity. More waves allows for greater stretch.

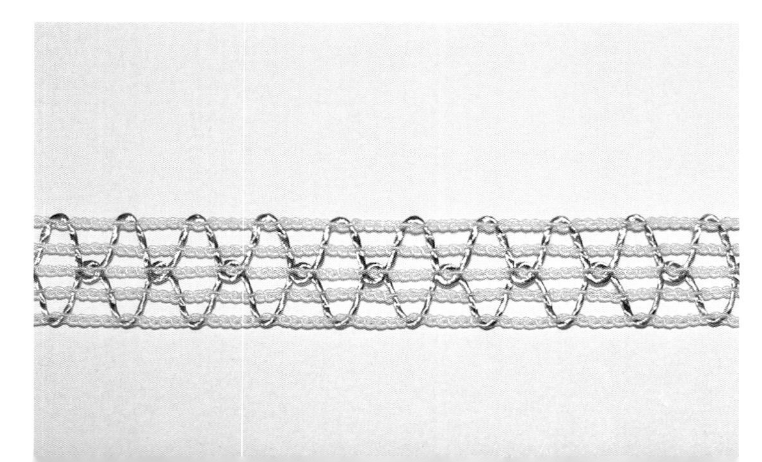

CONDUCTIVE CLAD FIBRES

Metal clad fibres have been developed as an alternative to standard solid metal wires. Each thin fibre is composed of a Polymer core, which is then coated with a metallic layer. These fibres are then coiled together into bundles which form light, supple cables combining the lightness, strength and flexibility of textiles with the conductivity of metals. The fibres made by AmberStrand® are Zylon™ fibres clad in copper, gold, silver or nickel.

Available in a variety of strand widths, from fine to heavy, the fibres have many potential applications. For example, as they are 80 per cent lighter than equivalent wires, use in aircraft wiring has the potential to save considerable weight and therefore fuel consumption and overall costs.

The wire can be coated in insulating material, making it great for applications where high temperatures, large strains and small spaces would impede regular copper wiring. It can also be used to incorporate electrical current into traditional textiles, particularly in applications where it is important for the wiring to be hidden.

PROPERTIES

Breaking strength between 4.13 and 20.50 kilograms or 9.1 and 46.2 pounds

Conductivity can be tailored

Double breaking strength of beryllium copper wire

Tensile strength minimum 590,000 psi

APPLICATIONS

Aerospace

Costume

Mechanical

INFO

www.metalcladfibers.com

www.glenair.com

Conductive thread was used as both a practical way to wire Diffus and The Danish Design School's Climate Dress and also a decorative detail. The dress' embroidery is embedded with conductive thread.

Images courtesy Diffus Design, The Danish Design School and the Alexandra Institute, © Michel Guglielmi.

INFLATED STEEL

OSKAR ZIETA

Polish designer Oskar Zieta makes ladders, chairs and stools—though not in the traditional sense. Although the designs look to be modeled on balloon contortions, they are light-weight yet strong, and able to support ten times their own weight.

Zieta's coat stand, for example, uses three rolls of thin stainless steel sheets. Once layered, the edges are then sealed with a laser welder. By pumping pressurised air through small valves the metal transforms into a coat stand. This "controlled loss of control" is the process that ensures uniqueness of each piece.

As a trained architect, Zieta developed FiDU (Free Inner Pressure Deformation) while developing techniques to stabilise sheet metal at ETH Zurich. Zieta explains that they "predict how the deformed object will look by the geometry of the cut, the pressure level, and the anisotropy [innate directional properties] of the material". The overall effect can never be fully determined beforehand as FiDU does not use a mould. In the future, however, he wishes to narrow the degree of difference post-inflation. Zieta's plans are skyrocketing: "My dream project is constructions in space, where FiDU's ultra-lightness and possibilities of volumetric expansion could be crucial." His team is testing the technique with metals like aluminium and copper in order to make a bridge and ultra-light-weight wind turbines and are only just beginning to explore the architectural possibilities of the material.

OPPOSITE
Zieta's range of functional seating, Plopp, in stainless steel. The detail shows the laser-welded edges and inflated form.

LEFT
Zieta's new work exploring the process in architectural applications. The Architonic Concept Stage III is a project devised by Architonic to discover whether FiDU technology might allow temporary structures, able to be transported flat and inflated in location.

BELOW
A prototype FiDU bridge form.

LIQUID METAL

PROPERTIES

Amorphous

Brittle

Good compression and tensile strength

APPLICATIONS

Industrial coatings

Medical

Military

Sports

INFO

liquidmetal.com

www.nasa.gov

Liquid Metal is a completely newly structured alloy capable of turning from a liquid structure or non-crystalline to a solid at room temperature, without having to go through rapid cooling.

This new process created a completely new kind of alloy capable of lifting 136 metric tonnes or 300,000 pounds on a one-inch diameter bar, as compared to titanium, which can lift 80 metric tonnes or 175,000 pounds in a comparable size. It is heralded as the biggest breakthrough in materials since thermoplastics. Although it has good tensile and compression strength, it is extremely brittle and can shatter easily.

Head brought out their Liquid Metal Radical in 2003, which won Business Week's 25 best products that year. Other sporting goods also use Liquid Metal, such as Rawlings baseball bats.

BELOW
A liquid metal sample.

OPPOSITE
Liquid metal is currently added to a number of everyday electronics products. It can be easily formed and has good compression and tensile strength, making such products very hard wearing. Images © Liquidmetal.

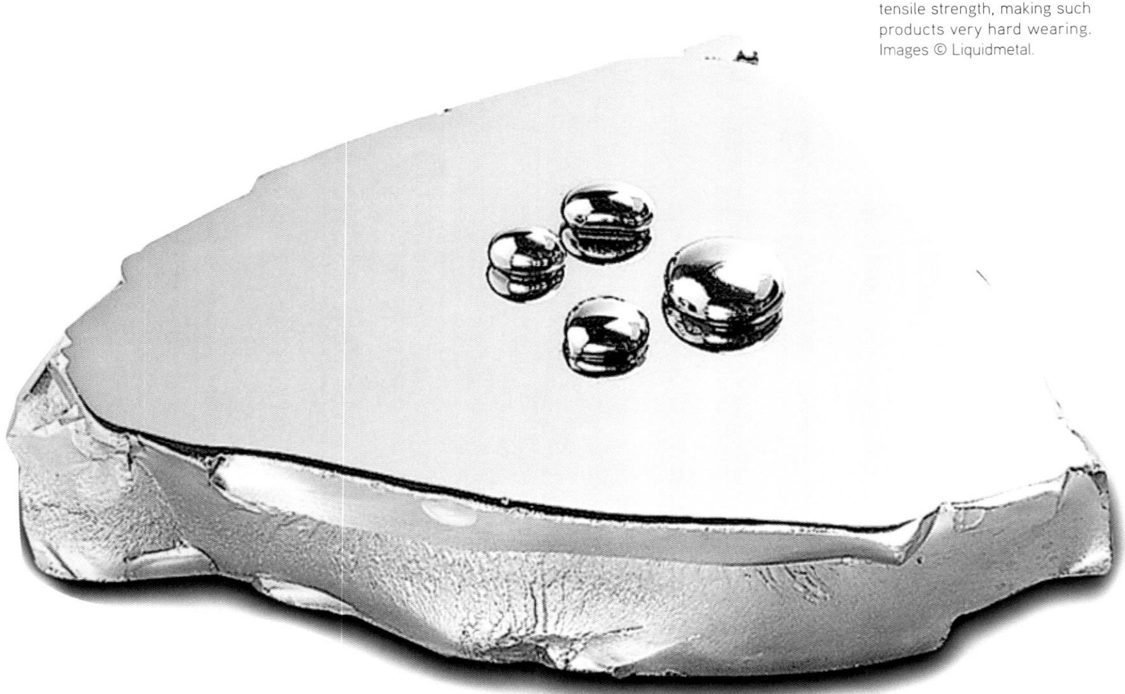

SONY
MENU
SCENE
LIQUIDMETAL

SINGLE CRYSTAL NICKEL SUPERALLOYS

PROPERTIES

Withstands huge stress

Can function at a higher temperature than previous blades

APPLICATIONS

Jet engines

INFO

www.synl.ac.cn

www.biam.com

Turbine blades in aircraft jet engines have to withstand enormous pressure and heat, and therefore have often suffered in the past from creep, fatigue, thermo-technical failure and hot corrosion. Replacing turbine engines is extremely costly to airline companies. Therefore, since the 1940s, much research has been done on the production of superalloys for turbine blades, with sophisticated material advances improving the overall performance of the jet engines, enabling us to fly higher, further and faster.

Single Crystal Nickel Superalloys are the latest development, superseding even the best superalloys of the past. When a solid metal forms, the arrangement of its constituent crystalline grains determines its strength. Single Crystal Alloys are made by isolating a single vertical crystal using a spiral selector as the liquid metal is poured into a mould. The metal goes through a process of controlled cooling which enables microstructural control of the newly forming solid, thereby developing what is known as a "Single Crystal Alloy", where only one metal crystal is allowed to grow into the mould of the turbine blade. A process called hot isostatic pressing eliminates pores, further strengthening the blade.

Looking at a turbine blade with the naked eye, before it has had its spiral sprig cut off ready for insertion into an engine, it is actually possible to see the alignment of the crystals, their jostling thoughout the spiral and then the triumph of a single crystal emerging out of the other side of the spiral into the solidified form of the blade itself.

BELOW
A jet engine turbine blade with its spiral selector still attached. The completed blade, which is the area to the left of the spiral, is comprised of only one crystal.

OPPOSITE
The view into a jet engine turbine. Jet engines are, in fact, comprised of hundreds of these Single Crystal turbine blades, which are precisely curved to draw the air through, thus propelling the plane at great speeds through the air.

THERMO-BIMETALS

PROPERTIES

Smart material

Heat-sensitive

APPLICATIONS

Thermometers

Product design

Architecture

INFO

www.dosu-arch.com

Thermo-bimetals are new self-actuating responsive materials that are comprised of sheets of differing metal alloys laminated together. When two metals which expand at different rates when heated are joined together, the structure that they form will bend as the metals fight each other into delicate contortions. If the composition of a Thermo-bimetallic structure is carefully controlled these contortions can provide a useful and embedded structural response.

Doris Sung at the University of Southern California has been researching the use of these materials as cladding for buildings, with the potential to initiate ventilation automatically based on the difference in temperature between the interior and exterior. The research will lead to the creation of facade panels, blocks or tiles that effectively allow a building to breathe. Such tiles also have potential to be used as a safety device, shutting the air out of a building in case of fire.

BELOW
A heat map study demonstrating the way the tiles will heat up over the course of a day from two views.

OPPOSITE LEFT
Three details taken at different times of the day illustrating the subtle opening of the tiles. When exposed to the sun's heat the thin Thermo-bimetal tiles gently lift outwards, allowing a breeze to enter the form.

OPPOSITE TOP RIGHT
A CAD sketch of the final form.

OPPOSITE BOTTOM RIGHT
The realised shelter, demonstrating the use of the materials within a smart structure, capable of altering the amount of shade it gives and breeze it permits into the space it creates. Application&Materials Gallery, Los Angeles. Images courtesy Doris Sung.

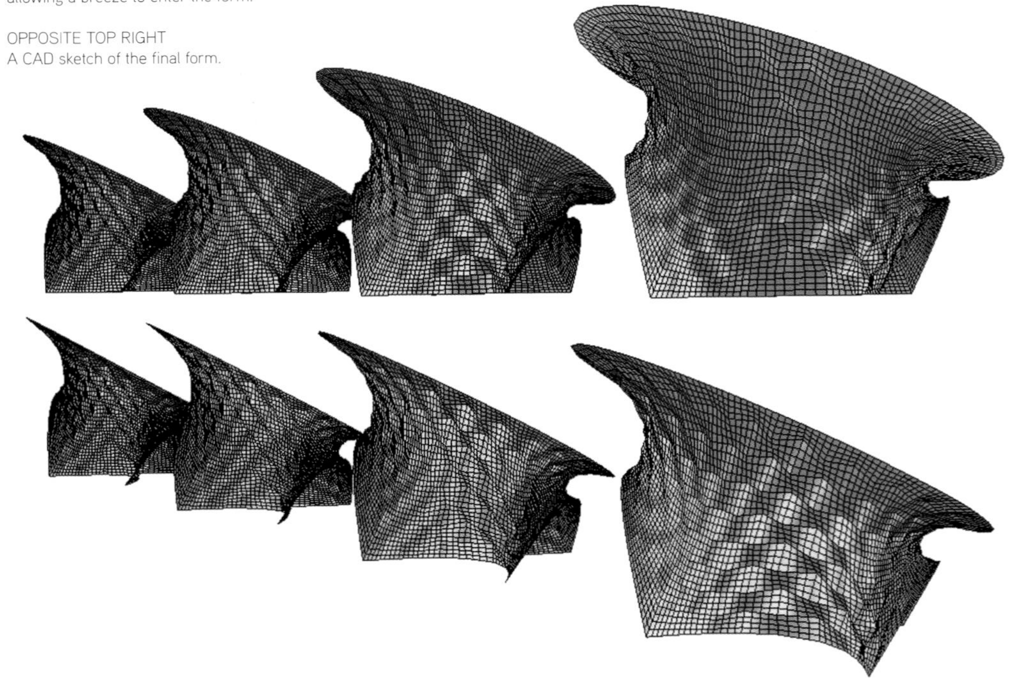

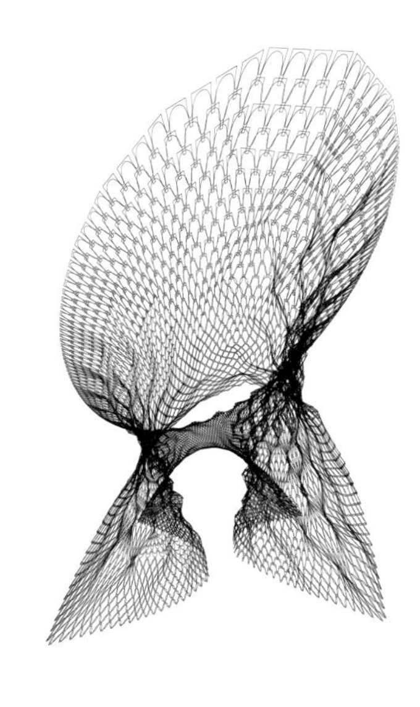

MICROLATTICE

PROPERTIES

High strength

Recovers shape after compression
99.9 per cent air 10 per cent
less dense than the lightest
aerogels —0.9mg/cm³

APPLICATIONS

Aircraft componets

Shock absorbers

INFO

www.hrl.com

This metal Microlattice toppled aerogel from its podium in 2011 to become the "lightest solid on Earth". It is comprised of very small tubes arranged in a Microlattice formation, the walls of which are only 100 nanometres in diameter.

The structure is produced using a templating method, where a nickel–phosphorous alloy is deposited upon a Polymer 'scaffold'. The polymer is then etched away leaving the metallic framework behind. In this form, the nickel–phosphorous alloy exhibits markedly different properties to its bulk counterpart. Nickel-phosphorus is normally quite brittle, but when arranged in the Microlattice structure the propagation of cracks is severely restricted, resulting in a final form which is much more elastic.

This metallic Microlattice was produced by a research team led by Tobias Schaedler of HRL Laboratories in Malibu. He writes, "by changing the structure at these levels you get completely different properties from the bulk material, which is a very powerful concept.... We think we may be able to introduce new properties into materials by this kind of structured porosity."

The Microlattice is so light it can rest on a dandilion without breaking it. Image © HRL Laboratories, LLC, photograph by Dan Little.

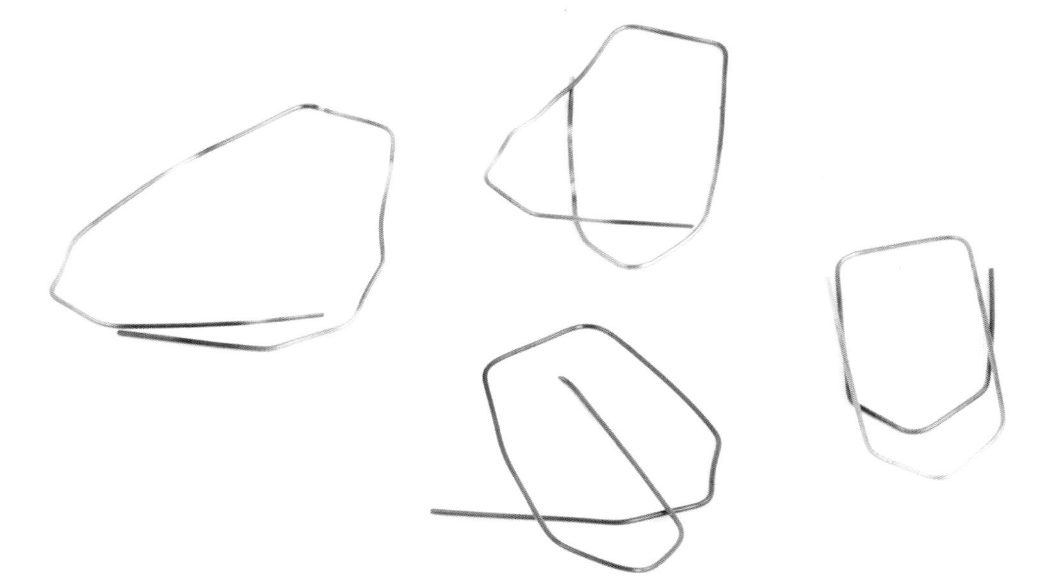

SHAPE MEMORY ALLOYS

Shape Memory Alloys are a general class of smart metals capable of remembering their shapes. Once deformed, heating them above their transformation temperature or giving them an electrical charge will trigger them to return to their original shape.

Shape Memory Alloys have been around since the 1930s and are in all sorts of everyday items, such as kettles. They are the element that turns an electric kettle off when the water boils, taking advantage of the simple action of pressing a button to deform the spring and allowing the heat to reform it. Very effective in a kettle and even more effective in heart valves and actuators, these metals are only just beginning to be used for art and design purposes whereas they have traditionally been used in industrial and medical applications.

There are a few different varieties, from metals that only remember one shape to metals capable of remembering two shapes and that are triggered by different temperatures.

PROPERTIES

Smart material

Can act as an automatic switch

Engineered for different properties

APPLICATIONS

Responsive tools

Switches

Actuators

INFO

store.migamotors.com

www.mindsetsonline.co.uk

ABOVE
A series of Shape Memory Alloys deformed and returning to their original shape. They have been bent out of shape and each successive image represents more heat applied to the material.

LEFT
These are the Shape Memory Alloys in the common electric kettle. Pushing the kettle's button compresses the metal. The metal decompresses when the water has boiled, turning the kettle off.

CASE STUDY

REACTIVE ENVIRONMENTS

HYLOZOIC GROUND

Hylozoic Ground was a retrospective exhibition in 2011 that celebrated the collaborative work of architect and sculptor Philip Beesley and his team. The extraordinary display embodied recent innovations in architecture, involving mechatronics, synthetic biology and art. The exhibition took the form of a fake forest made out of tiny acrylic meshwork links, patterned and shaped to resemble the fauna of a woodland setting. These lattices were covered with a layer of mechanical fronds and filters to produce an intricate network of immense complexity.

Hylozoic Ground is an immersive creation that, like a living entity, moves and responds around the passage of its visitors. Next-generation artificial intelligence and interactive technology create an environment that is, appropriately for its title, almost alive. Using touch sensors and Shape Memory Alloy actuators for their kinetic properties, the whole installation flutters and responds to human interaction.

Hylozoic Ground was chosen to represent Beesley's native Canada at the 2010 Venice Biennale in Architecture, the most important and prestigious international event in contemporary architecture, thereby bringing Shape Memory Metals to an international audience in a new context to create a stunning sculptural and responsive environment.

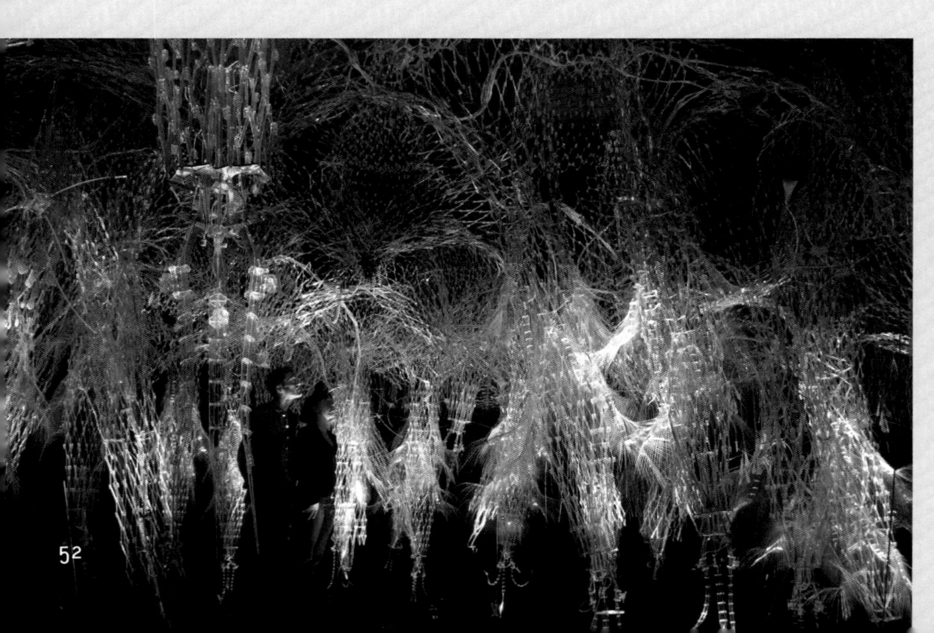

LEFT
Two figures walking through the instalation, demonstrating the immersive nature and scale of Hylozoic Ground.

OPPOSITE
Details showing the fronds and feather-like tendrils, which are controlled by shape memory actuators. All photographs © Pierre Charron.

NITINOL

PROPERTIES

Superelasticity

Shape memory

Highly biocompatible

APPLICATIONS

Heart stents

Medical wires

Underwire for bras

INFO

www.nitinol.com

www.nitisurgical.com

memry.com

Nitinol is an alloy of titanium and nickel that is often simply referred to as a Shape Memory Alloy. The metal is easily bent into many shapes but when heated to a specific temperature, it transforms back to the shape it was originally made into.

Nitinol also exhibits high elasticity and biocompatibility, which, when combined with its shape memory properties, make it the perfect material for stents—the little valves designed to keep human arteries open, preventing heart attack and strokes. These chilled stents are designed for permanent implantation. Once inside the blood vessel and adjusted to body temperature they spring open, expanding to their designed size and holding the plaque against the arterial wall. The stent is vital for reinforcing weak vein walls and widening veins that are too narrow.

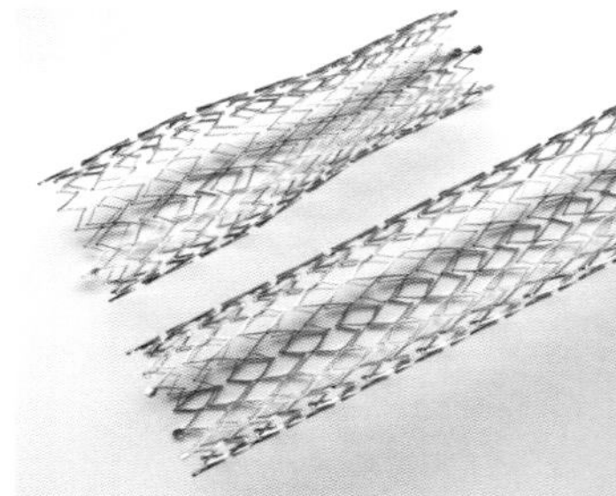

TOP LEFT
A vena cava stent made from Nitinol, designed to be compressed before insertion, whereupon the heat from the patient's body opens the stent, widening the valve.

TOP RIGHT
A partially compressed stent demonstrates the differing diameters needed for implantation.

BOTTOM
Small heart stents. Each of the stents pictured is under five millimetres in diameter, demonstrating how small and intricate the stents must be to function inside a delicate human vein. Images courtesy memry.com

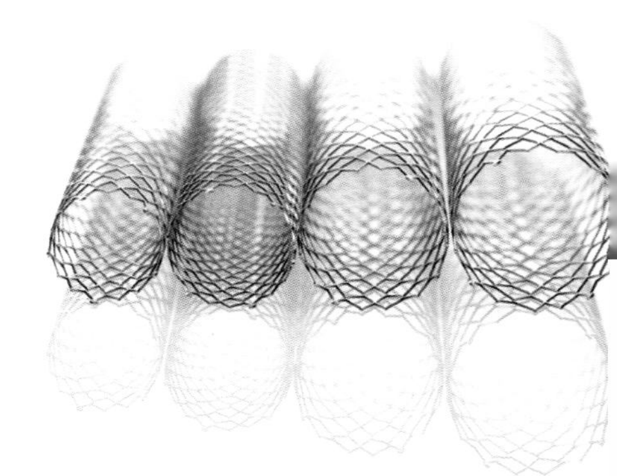

MAGNETIC SHAPE MEMORY ALLOYS

Once deformed, Magnetic Shape Memory Alloys (also known as "ferritic magnetic alloys") revert back to their precise original shape when exposed to a magnetic field. For example, after undergoing compression, exposure to an electromagnet would see a material elongate and regain its original form.

The shape change that Magnetic Shape Memory Alloys undergo occurs very rapidly and precisely, making them perfect for high-tech applications such as robotic arms, which require exact positional control. For example, researchers at the University of the Basque Country have used them to develop prototype robotic arms capable of precision tasks in medicine and engineering, which require actions to be performed to accuracies measured in nanometres.

PROPERTIES

Exhibits Villari (inverse magnetostrictive) effect

Controlled spring properties

Magnetic Shape Memory Effect—up to 6 per cent elongation in a magnetic field

APPLICATIONS

Vibration dampers

Sensors

Energy harvesters

Engineering

INFO

www.goodfellow.com

www.ikasketak.ehu.es

MAGNETIC SHAPE MEMORY FOAM

Developed using a Magnetic Shape Memory Alloy of nickel, manganese and gallium, researchers from Northwestern and Boise Universities have produced a Magnetic Shape Memory Foam. Its main benefit is the reduced mass of components compared to their solid counterparts, which results in lighter and lower costs for memory precision components.

The alloy exhibits new characteristics for Shape Memory Alloys, as it changes shape when exposed to a magnetic field, retains this shape when the magnetic field is removed, but returns to its original position when rotated 90 degrees to the direction of the magnetic field. The result is potentially a tiny linear actuator that could replace similar more costly devices made from multiple parts.

Introducing small pores into its already porous structure enhances the foam's abilities even further.

PROPERTIES

Reliable

Can be repeatedly activated with no reduction in movement

APPLICATIONS

Fast operating actuators

Sonar devices

Electromagnetic sensors

Microbotics

INFO

news.boisestate.edu

www.nsf.gov

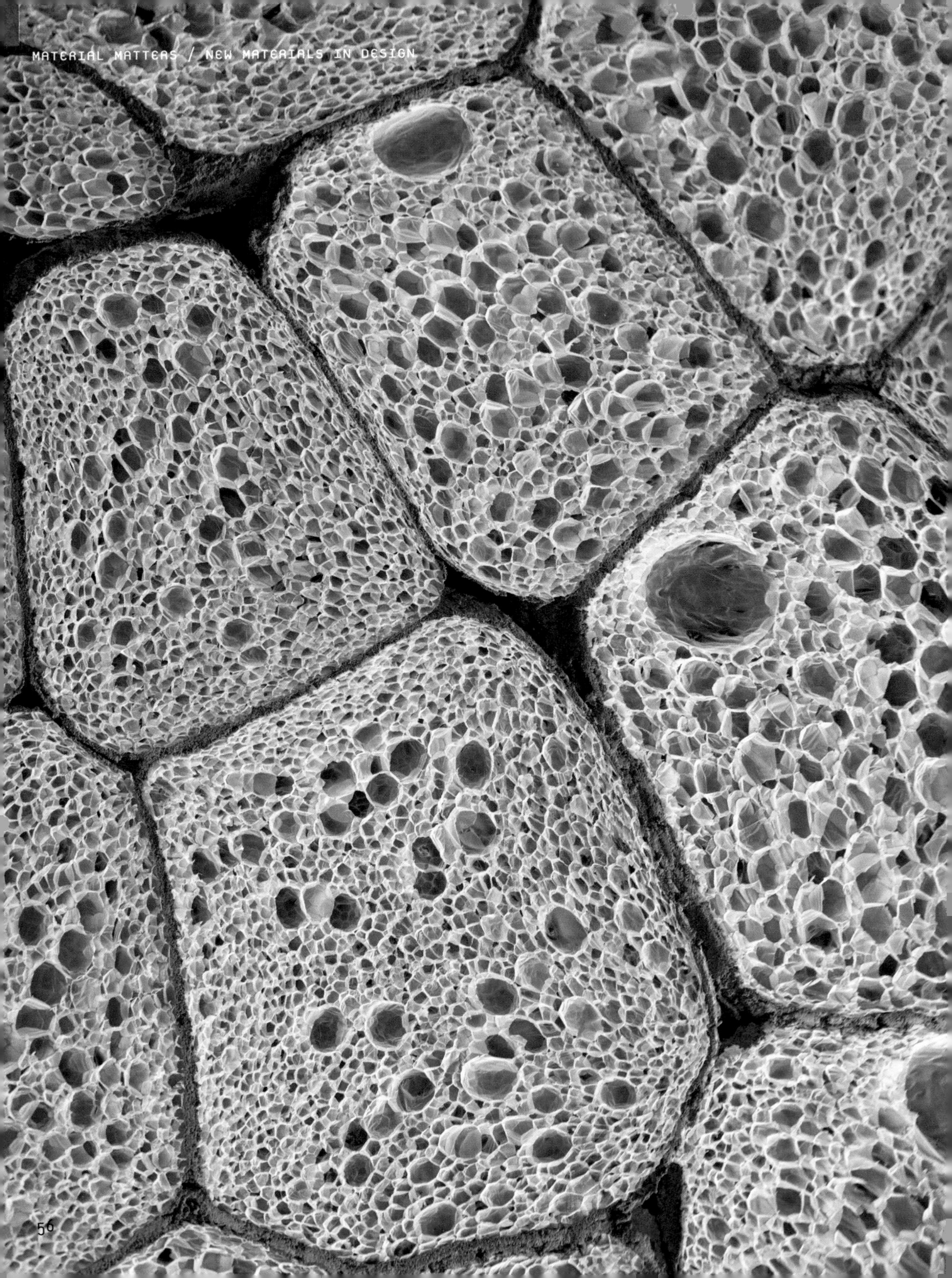

METAL FOAM

Metal Foams combine the advantages of metals with the structure of either open or closed-cell forms, and are useful for a range of experimental applications from orthopaedics to automobiles, where the light-weight structures are being experimented with as possible crumple-zone materials.

Foamet™ is an open-celled Metal Foam made from a powder-metallurgical manufacturing process. Using this process allows Foamet™ to be made out of almost any sinterable metal powder. Closed-cell foams more commonly create bubbles in the metal by a process of gas injection and result in a non-porous material that can float.

PROPERTIES

High permeability

Low specific density

Cell diameter from 0.3–5 mm

Porosity between 70 and 95 per cent

Adjustable inner surface up to 4000m² per m³

APPLICATIONS

Crumple-zone materials

Automobiles

INFO

www.hollomet.com

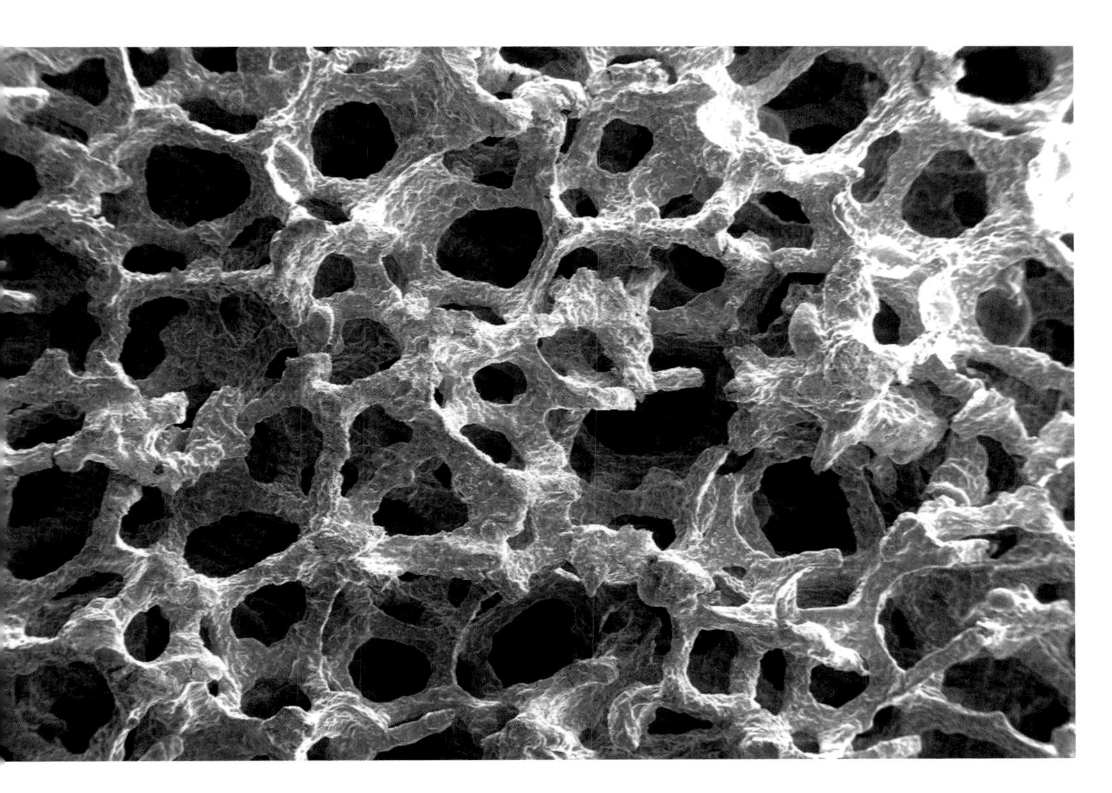

OPPOSITE
Metal foam. Coloured scanning electron microscope (SEM) image of foamed nickel metal, which results in a larger surface area for a lower overall weight. The foam is most commonly used in the battery industry. Image © Science Photo Library.

ABOVE
A microscope image of Metal Foam. The structure of the foam is formed by sintering powdered metals, creating a fine texture.

TITANIUM FOAM

PROPERTIES

Easy bonding with bones

Flexible

Low corrosion

APPLICATIONS

Biomedical implants

Replaces damaged or broken bones

INFO
www.fraunhofer.de

Fraunhofer, the German industrial and medical research firm, have developed a Titanium Foam that is designed to replace damaged or broken bones. As it is made of an open foam structure it encourages the existing bone cells to grow through the matrix, thus becoming incorporated into the skeletal structure.

Titanium has long been used for medical implants because of its low corrosion and easy bonding with bones. However, the use of titanium rods had caused problems because the metal was stiffer than the human bone, resulting in a mechanical miss-match and possible fractures. By inserting foam, rather than solid metal, there are two advantages; the body can grow through the structure, ensuring a strong bond, and the foam is flexible enough that it can cope with load-bearing activities in a similar way to human bone.

TiFoam, as it is branded, is made by filling open-cell polyurethane foam with very fine titanium powder and a binding agent. Once the titanium has bonded the polyurethane is vapourised. The foam is actually a cast of foam, rather than a foamed metal, created in the way that a Polymer foam might be. The residual titanium is further heated and compressed until it resembles the spongiosa structure of human bone. In fact, the implant begins to bind with bone immediately and patients are encouraged to use it right away because doing so increases the growth through the titanium and therefore strengthens the overall structure.

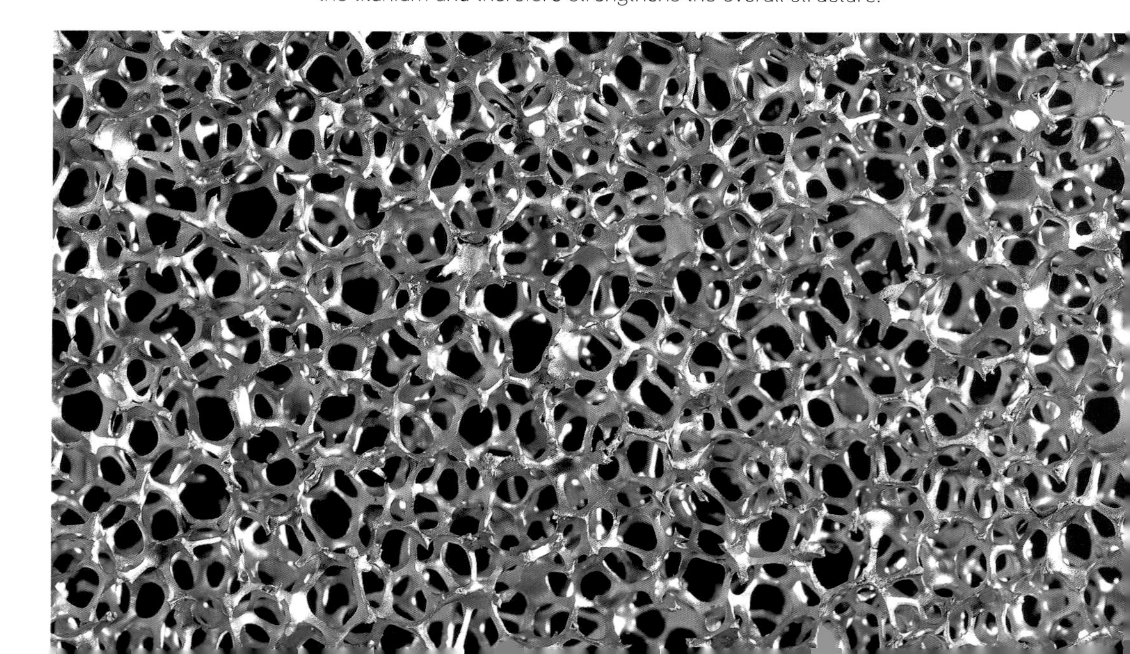

Detail of the Titanium Foam developed by the Fraunhofer Institute in Germany. The structure of the material so closely resembles bone that the bone is able to not only bond to the metal but grow through it, making it extremely biocompatible. Image © Fraunhofer Institute.

HOLLOW SPHERE STRUCTURES

Hollow Sphere Structures are a type of Metal Foam with adjustable and engineered porosity formed around spherical carriers. The high surface area and round structures mean that it has an extremely high strength to weight ratio.

Hollomet manufacture a product called "Globomet." It is formed by coating a round polystyrene ball with fine layers of powdered metal. The powder is then bonded by sintering. The polystyrene is burned away during the process, leaving only a porous layer of fused metal around, hence the name "Hollow Sphere." The wall thicknesses can be specified to be between a few tenths of a millimetre to one millimetre thick.

PROPERTIES

Diameter 1.5–10 mm

Wall thickness >20 µm

Can be processed using traditional tools 40 to 70 per cent lighter than solids

APPLICATIONS

Aircraft

Bearings

Lighter replacement for current metal applications

INFO

www.hollomet.com

www.fraunhofer.de

LEFT
A detail magnified to show the fine texture of the sintered metal on the outside of each of the 'Hollow Spheres.'

BELOW
A selection of filtration materials made from Hollow Sphere Structures. Each hollow sphere is formed by sintering and each sphere is attached to the others through the sintering process, making the shapes very robust. Images courtesy Hollomet™.

GLASSES

Glass is a material of extremes. It exhibits strength and endurance, is extremely stable, does not change appreciably over time, is resistant to chemical attack and is almost totally inert. On top of this, it is magnificently transparent and can be engineered to produce a wide range of optical effects. However, if over-stressed it does not bend or deform, but shatters beyond repair, revealing its underlying brittle fragility.

The fundamental building block of glass is silica. Pure silica glass is composed of an ever repeating combination of pyramidal silicon oxide. The oxygen atoms prefer to connect with two silicon atoms, so when another molecule comes along, the two fuse, with the oxygen providing a bridge between the silicon atoms. In a hot molten glass, there are many millions of these silicon oxide units floating around, freely making and breaking connections with each other, and it is the transience of these connections which allows the glass to flow. However, as it is cooled, these connections become increasingly hard to break, making the liquid glass thicker and more viscose. As the temperature continues to drop, these connections start to form permanently, fixing the atoms in position. The formation into which these building blocks are arranged is random—there is no repeating order or nice tidy packing structure like we see in crystalline materials. This makes the structure amorphous, a defining feature of all glasses.

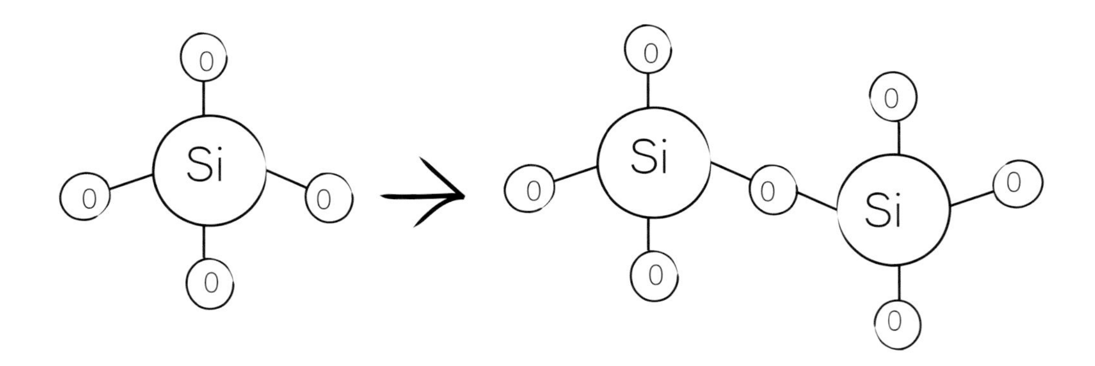

ABOVE
A diagram illustrating the molecular bonds formed in glass. Each silicon atom bonds to four oxygen atoms, a motif that is repeated throughout the solid.

PREVIOUS PAGE
Detail of microscopic glass beads.

Most glasses are not purely silica, with other ingredients being are added to tune properties and behaviour. For example, adding metal oxides to silica yields soda lime glass. The metal atoms disrupt connections between the silicon and oxygen which lowers the melting point, allowing the glass to be worked at much lower temperatures, a key factor in producing glass on an economically viable scale. Another example is Borosilicate Glass, made by adding boron oxide, resulting in a glass with excellent temperature resistant properties. However, it is not just the starting ingredients that define the properties of a glass—the manufacturing process is hugely influential too. Glasses can be pressed, blow moulded, drawn, rolled and cast, and they must be cooled in a very controlled manner in order to avoid unwanted stresses forming in the structure. It is this fine-tuning of ingredients and manufacturing techniques that give rise to the vast array of glasses we have today.

PH and ZL

A detail of a lighthouse Fresnel lens. The shape uses glass to maximise the light emitted from the lighthouse using the least possible material for the greatest possible effect. Today these complex shapes use CNC milling technology.

One common myth surrounding glass is that it flows very slowly over time, thus resulting in the glass in old windows being thicker at the bottom than at the top. This is not the case. Once solid, glass is fixed and does not creep, flow or shift shape. The variability in thickness of old windows in fact arises from the conditions of manufacture, where it was extremely difficult to obtain uniform sheets.

Everyday applications of glass are evident in windows, lenses, light bulbs and bottles. Different applications utilise different properties, and the many advances in glass production give rise to less obvious manifestations of the material: toughened glass, suitable for tough environments; Fibreglass, where glass fibres are mixed with different polymers to produce tough and light-weight structures; or Optic Fibre cables, where long lengths of flexible glass fibres are used to transmit light and carry data. In most uses of Glass it is the optical properties that are most highly prized, as has been the case for many centuries. For example, pieces of Egyptian jewellery have been found where beads of fulgurite glass (produced as the result of a lightening bolt striking the desert sand) have been placed at the heart of their settings, celebrating the material above gold or precious stones.

PH and ZL

LEFT
A sample of fulgurite—natural glass created when lightening strikes sand and fuses it to glass.

OPPOSITE
The glass ceiling of The British Museum demonstrates the latest developments in glass for architecture. Each individual triangle is a different size and form in order to accommodate the curve of the overall roof form.

CASE STUDY

TURNING SAND TO GLASS

THE SOLAR SINTER

Trialled in the Sahara Desert, the Solar Sinter is essentially a 3-D printer that uses the sun's energy to transform sand into glass. Designed and built by Markus Kayser whilst studying at the Royal College of Art, it was a development of a semi-automated Sun-Cutter that he operated in the Egyptian Desert.

Sintering involves heating powders to form solids. In the case of the Solar Sinter, the sunlight is the raw energy, and sand the raw material. The silica sand becomes molten when heated to its melting point, and then solidifies as glass.

The Solar Sinter, though driven electronically, works at a measured pace. A computer is used to draw up images of the final structure, then encoded files are fed into the sinter which then starts shifting the sand box, via aluminium frames, to the correct coordinates. A special lens focuses the sun, heating the point of contact with the sand up to 1600 degrees Celsius or 2912 degrees Fahrenheit, with a sun-tracking device always ensuring that the lens remains facing the sun. Upon completion, the objects are rough and

LEFT
The Solar Sinter in the desert. The machine can be folded up and transported. Using only energy from the sun it is able to transform the desert sand into glass.

BELOW
A detail of the Solar Sinter showing the tunnel of lenses used to concentrate the sun's rays. The low-tech lenses demonstrate the immense power of the sun. With just a few lenses, sand can be heated to temperatures high enough to melt it.

OPPOSITE
A glass bowl made from the sinter process with a bright sun 'laser' spot on the sand beside it, concentrated to almost the white hot temperature required to melt the sand. Images courtesy Markus Kayser.

sandy on the reverse side but solid glass on the top. The colour of the glass reflects the sand grains used to build it, therefore every desert will yield unique compositions.

In the future, Kayser envisages using the desert floor as his canvas and hopes to print straight onto the sand. If lenses are used to melt sand into wall strucutres, the prospect of solar-sintering on a larger scale could lead to what he calls "a new desert-based architecture". Kayser adds, "this image of a multiplicity of machines working in a natural cycle from dusk till dawn presents a new idea of what manufacturing could be".

Still in its infancy, the project has a long way to go before influencing current production models, or building walls. However, it already makes us reconsider the power of the sun for industrial processes, and allows us to reconnect with the wonders of glass as a made material, derived from silica and used across all sorts of high-tech and domestic applications, from the dainty and decorative to objects of extreme optical precision.

SUPER THIN ECO GLASS

PROPERTIES

Thickness—0.03mm–1.1mm

Coefficient of expansion close to ceramic

High chemical resistance

Can be customised by shape and size to order

APPLICATIONS

Screens

Microscope slides

INFO

www.schott.com

www.acg.com

Most of the panes of glass that we come into contact with are produced using the float method. The thinnest float glass achieved to date is 0.1 mm thick—about the thickness of a piece of paper—produced by ACG in Japan. However, display screens and touch screens often require thinner panes.

To make thinner glass panes the overflow downdraw method can be used. Here, molten glass is continually poured into a trough until it overflows. The overflow is pulled off the flat edges in long, extremely thin flat sheets. The advantage of this method is that the glass surface does not come into contact with any other materials, such as the tin in the float glass process, and can be made extremely clear with a fire polish. Gravity, the speed of the pour and pull of the mechanical arms control the thickness of the glass, enabling the thinnest glass screen material yet—Schott's D363T, which can be produced at only 0.03 millimetres.

Schott's D263T is a colourless Borosilicate Glass made without heavy metals, hence its 'eco' credentials. Eliminating heavy metals makes the manufacture of the glass, including its waste production, less harmful to the environment.

LARGE-SCALE SHEET GLASS

PROPERTIES

Laminated

Tempered

XL glass can measure up to 8 x 3.3m

APPLICATIONS

Architecture

INFO

www.glassolutions.eu

www.xxlglass.net

www.eckelt.at/en

Making glass panels larger does not necessarily require a change in the composition of the glass or new production processes, rather it simply requires up-scaling the mechanisms of production. To make these tempered and laminated window panels measuring almost six metres or 19.7 feet, as used by Perkins Eastman Architects for 15 Union Square West in New York, the glass manufacturers Saint-Gobain had to build a new super-plant in Austria, to cope with the scale of production. The glass can be used to glaze double-height atrium spaces seamlessly and make facades appear more uniform, with fewer joins and fixings.

The largest seamless panes of glass in the world are installed in the Apple Store in Sydney, measuring ten metres high.

INSULATED GLAZING

Insulated Glazing, for example triple-glazed or noble gas-filled double panels, can act as extremely efficient insulation for buildings.

By filling double-glazed panels with noble gases such as argon, krypton or xenon, instead of dehydrated air, heat convection through the panel is greatly inhibited. Noble gases are extremely inert, which means they do not react easily, and they do not transfer heat readily.

Triple-glazing is the norm in Scandinavia, where outdoor temperatures regularly plummet far below zero. Iplus city E is a triple-glazed panel that is not only highly thermally protective, but also extremely good at insulating against noise. It also has the added bonus of being very tough.

PROPERTIES

Triple glazed: Ug=1.1 W/(m²K)

Suitable for thermal toughening

Argon filled: 34 per cent lower thermal conductivity than air

Available with additional coating for heat and sun protection

Environmentally friendly

Sound insulation value up to 40dB

APPLICATIONS

Architecture

INFO

www.pilkington.com

www.glasstecwindows.co.uk

www.interpane.com

www.jma.it

90 per cent Argon Gas inserted in 18 millimetres spacing between the exterior and interior glass sheets of the Lake Lugano House by JM ARCHITECTURE.

INFLATED GLASS

PROPERTIES

Elastic when heated

Fragile

APPLICATIONS

Sculpture

INFO

www.urbanglass.org

www.zanebennettgallery.com

This novel use of glass shows how an artist's understanding of the mechanical properties of glass and a spirit of play can lead to new and innovative structures. In a process similar to that of industrial blow moulding, Matthew Szosz, a glass artist, developed a method of inflating fused panes of flat window glass to create glass sculptures unlike anything seen before. By stacking panes of float glass, bonding them at the edges and then heating them in a kiln to a temperature at which the glass would be malleable, Szosz was able to inflate the glass envelopes using compressed air.

Recently he has developed a number of other sculptures using heat and gravity to form three-dimensional shapes from flat glass sheets manipulated when hot then allowed to cool into new solid forms that visualise the tension of a recently inflated balloon.

BELOW
Euplectella, glass and stainless steel wire, 2010. 3 x 2 x 1m. Float glass was heated and stretched using gravity once in its semi-molten state.

TOP AND BOTTOM RIGHT
Inflated Glass Sculpture, and *Untitled* (Inflated Glass), 2007. This new kind of material was formed from familiar float glass fused and inflated.

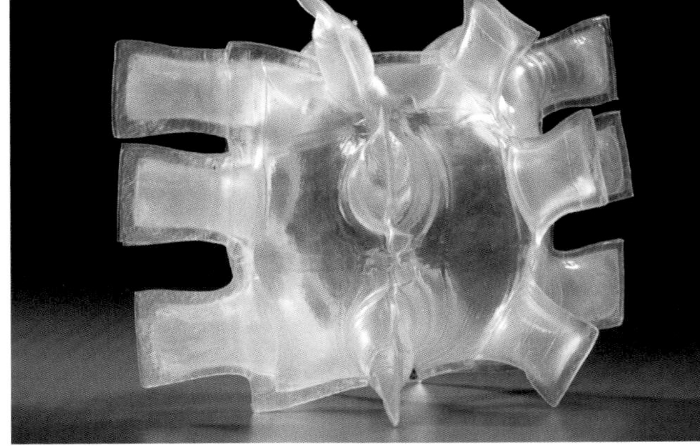

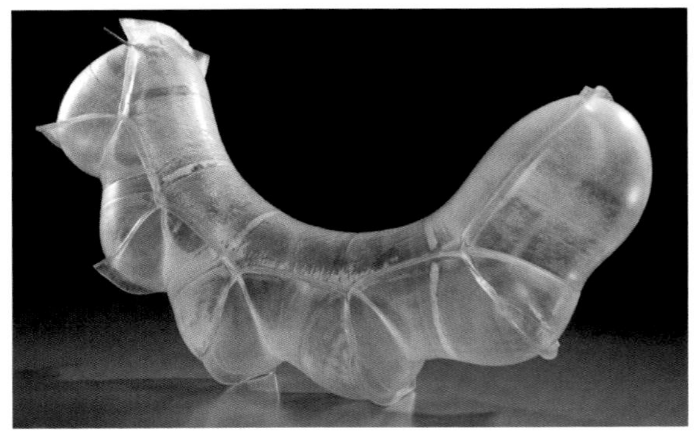

BOROSILICATE GLASS

Borosilicate Glass is made from silica doped with boron oxide. The inclusion of the boron creates a glass that is difficult to work, but exhibits extremely good heat resistant capabilities.

This glass was developed by Otto Schott in the late nineteenth century, with Corning glassworks introducing PYREX® into the marketplace in 1915. PYREX® can be used at extreme temperatures (oven, microwave or freezer) where normal soda-lime glass would crack and shatter due to excessive thermal expansion or contraction. This is why Borosilicate Glass is used to make laboratory glassware such as test tubes, flasks, beakers and distillation chambers.

Beyond its extensive use in laboratories, Borosilicate Glass has seen somewhat of a revival thanks to developments in glass art. Due to its high melting point it needs to be manipulated using welding equipment but can be used to create vast structures in large-scale installations.

PROPERTIES

Heat resistant

High melting temperature

Thermal expansion coefficient is about a third of soda-lime glass

Thermal shock resistance

APPLICATIONS

Sculpture

Laboratory glassware

INFO

www.corning.com

www.pyrex.com

www.schott.com

Susan Plum, *Woven Heaven, Tangled Earth*. Fine canes of Borosilicate Glass were heated using a welding torch until molten enough to fuse. The Borosilicate Glass' low expansion index makes this possible. Image © Corning Museum of Glass and the artist.

CASE STUDY

LIGHT AS MATERIAL

JAMES CARPENTER DESIGN ASSOCIATES

James Carpenter Design Associates has 30 years of architectural experience in manipulating the natural phenomena of light. JCDA assess the surroundings of a building, taking its environment as a starting point in order to guide the design and ultimately enhance public perception of place. Valuing the importance of light in architecture and glass as a major factor in this, their designs channel light "in transmission, reflection and refraction". Indeed, James Carpenter is often referred to as a glass technologist or light artist. He became a MacArthur Fellow in 2004 for his exceptional and enhanced creative work in architecture and design.

His studio nurtures the similarities found in different disciplines, pooling people from many professions such as architects, materials and structural engineers, environmental engineers and fabricators to work together. The studio uses glass, steel, wood and composites to complete a range of works, reaping the advantages of combining various skills. JCDA seek to "design projects that are poetic expressions of light and place". They re-establish a personal identification with light, a precious commodity in any urban setting of narrow

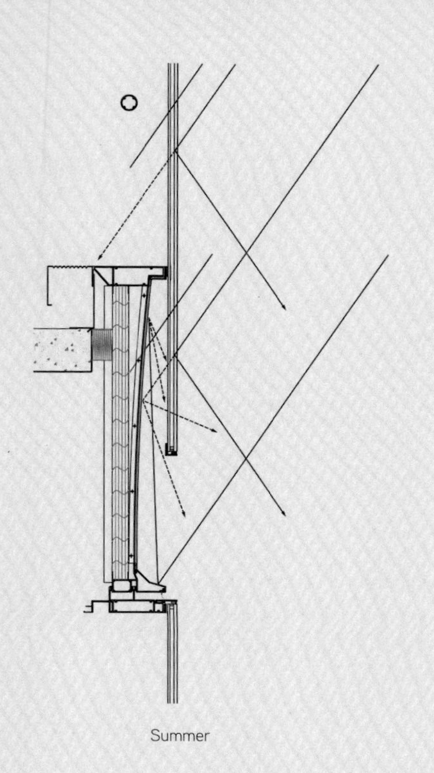

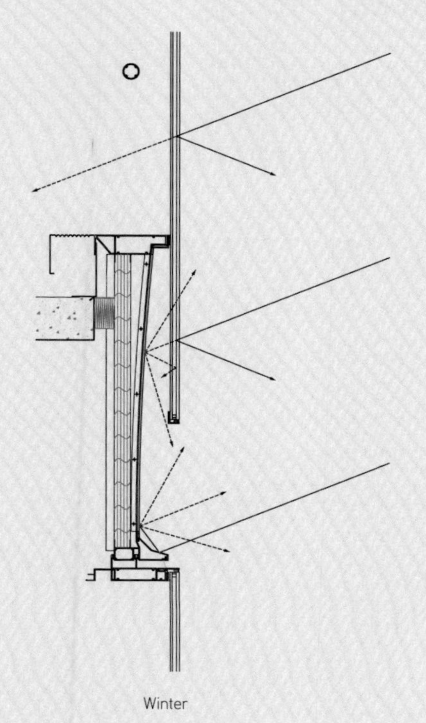

OPPOSITE LEFT
View up the glass facade of the building from pedestrian height. The glass makes the building appear to disappear into the blue of the sky.

OPPOSITE RIGHT
7 World Trade Center at dusk. The LEDs embedded in the base of the window panels take over from the job of the reflector, reflecting light up the window panes.

LEFT
Diagrams illustrating the way in which the reflector at the base of each window segment function. The light is reflected up through the overhanging glass panels, creating the impression of disappearance. Images courtesy James Carpenter Design Associates.

streets, skyscrapers and tall tower blocks. JCDA designed the first building to be rebuilt after 11 September, 2001. 7 World Trade Center is a parallelogram shaped tower comprised of 42 floors of offices.

While most Manhattan morning commuters see only high-rise buildings on the skyline, sunrise is a special time of day for 7 World Trade Center, when the design comes into its own. A specially shaped reflector catches and redirects the light, "transforming the quality of pedestrian experience at street level". At sunrise and sunset LEDs within the podium screen and lights in the lobby turn cobalt blue. The exceptionally transparent glass and acute corners produce the effect of the building disappearing into the sky at dawn and dusk. This response to both sunlight and shadow continually changes the colour of the building's front face, a facade as fleeting and transient as light. The lobby's interior contains a luminous ceiling volume. The podium wall lighting is white during daytime, changing into blue at the beginning and end of the day. The wall behind reception is a memorial to 9–11 of scrolling quotes; their reflection giving texture to the entrance.

TEMPERED GLASS

PROPERTIES

Compressive strength of over 100MPa in order to be classified as 'safety glass'

Tempered glass has a compressive strength of 69MPa

APPLICATIONS

Architecture

INFO

www.innoglass.com.my

www.tempered-glass.com.cn

The process of tempering increases strength and toughness, and in the case of glass is oddly counter-intuitive; glass can be made tougher by creating internal stress, which is what causes toughened glass to shatter so spectacularly into many tiny nugget-like pieces. The most common method for toughening glass used to be through a process of heating and cooling, but chemically treating glass has become more common and produces a tougher glass. The greater the surface stress exerted on the glass, the smaller the pieces of the shatter. Breaking into such tiny pieces reduces the risk of injury for the tiny nugget-like pieces behave more like a crumbling biscuit than sharp jagged shards of glass which can easily cause harm.

Thermal toughening is achieved by annealing, heating glass above a certain temperature and then cooling it rapidly, which allows the outside surfaces to harden while the inside remains molten. This exterior solidity creates tension and rigidity on the outside, but allows the interior to cool at a more moderate rate, creating internal stresses in the structure. Chemical heating is performed through a number of chemical processes, but the result is the same.

When a sheet of Tempered Glass is laminated and shatters it causes something called the "blanket effect", leaving a mat of shattered glass trapped in flexible clear laminate. This is also being used as a decorative effect.

SELF-CLEANING GLASS

PROPERTIES

Environmentally friendly

Low maintenance

Noise and solar control

Self-cleaning

Thermal insulation

APPLICATIONS

Building and construction industries

INFO

www.pilkington.com

www.selfcleaningglass.com

Self-cleaning Glass rids itself of dirt in a two-stage process. It is coated with a photocatalytic layer that harnesses the sun's energy to break down organic matter (i.e. dirt). It is also hydrophilic (or "water loving"), which means that rain water quickly spreads out into a thin film on its surface. This film of water runs off the window in a sheet, removing the broken down dirt as it goes, leaving behind a cleaned surface.

SUPER-STRONG GLASS

Gorilla® glass, made by the specialty glass designer and manufacturers Corning, is the toughest glass on the market today and is mainly used for screens of hand-held multimedia devices at risk of being scratched or even cracking in day-to-day use. The second generation of the glass is actually 20 per cent lighter and slimmer than the original, which makes it very promising for application in razor thin mobile devices.

The process used to manufacture Gorilla® glass stems from research done in the 1960s with the original intention of using the glass in car windshields to make them safer, but the cost of producing such large sheets was prohibitive. With the invention of smartphones, and the explosion of mobile touch screen technologies, Corning's material has found a new and widespread use.

Gorilla® glass is made from silicon dioxide aluminosilicate using an overflow downdraw process. Each sheet is just over half a millimetre thick as it is drawn out of the trough of molten glass, before being toughened using an ion exchange process. Here, the sodium ions in the aluminosilicate glass are replaced by potassium ions by submersion in a hot salt bath. As the potassium ions are larger than the sodium the density of the glass is increased, with a subsequent increase in toughness.

PROPERTIES

Light

Scratch resistant

Slim

Tough

APPLICATIONS

Hand-held multimedia devices

INFO

www.corninggorillaglass.com

A sheet of glass bending under weights demonstrating the flexibility of Corning® Gorilla® Glass. Image © Corning Incorporated, 2012. All rights reserved.

ELECTROCHROMIC GLASS

PROPERTIES

Automatically responds to light

Cost-effective

APPLICATIONS

Glazing systems

INFO

www.sageglass.com

www.smartglassinternational.com

Electrochromic Glass is capable of changing opacity when an electrical current is applied to it. Such glasses often contain active layers that have been engineered at the nanoscale and are at the cutting edge of glass technologies. Such layers often contain lithium ions that move from layer to layer, under electrical stimulation, orientating themselves to produce different optical effects. In practice, when a low voltage is applied to Electrochromic Glass, the material darkens, absorbing heat and blocking the glare from the sun. By reversing the polarity of the voltage applied, the glass turns optically clear again, so that on a cold or dark day it is able to optimise daylight and solar heat.

SageGlass, an American company, manufacture a specific type of Electrochromic Glass called SageGlass that alters the intensity of its colour when provoked by an electrical current. Such glass is being integrated into the glazing systems of buildings to enables the control of sunlight and glare without the need for window blinds—either at the touch of a button or as part of a detection system that automatically responds to the amount of light falling onto the building.

THERMOELECTRIC GLASS

PROPERTIES

Transparent

Slim

Don't require gas for heating

APPLICATIONS

Radiators

Heated windows

Hot plates/warmers

INFO

www.saint-gobain.co.uk

www.glassradiators.co.uk

Saint-Gobain, a glass manufacturer, make glass radiators with Thermoelectric Glass, which emits heat on application of an electric current. The glass is modified with a layer of conductive material that allows flat glass sheets to effectively conduct and emit heat. So far the technology has been used to create items such as 'invisible' radiators and heat emitting mirrors, which change our concept of traditional radiators as mono-functional devices. In a domestic environment, such innovations not only eradicate the need for unsightly radiators, but could increase the functionality of various other devices around the home.

THERMOCHROMIC GLASS

These glass tiles alter their colour in reaction to changes in temperature. The colour change originates from a thin thermochromic film attached to the back of the glass, rather than from within the glass itself.

The films are based on liquid crystal technology where, in reaction to heat, crystals rearrange themselves and change the colour of the film. They are particularly effective because the crystals move upon exposure to even small changes in temperature, where the temperature of a human body is enough to induce a colour change, as the image demonstrates.

As is crucial for the long-term use of such tiles, the crystals are able to repeatedly move back to their original arrangement, thus returning the tile to its original colour over and over again. Such durability is vital in architectural or decorative applications. An example of a practical use might be placing the tiles above an oven to provide a visual reminder if it is accidently left on.

PROPERTIES
Heat sensitive
Colourful

APPLICATIONS
Architecture
Decorative
Safety

INFO
www.inventables.com

Thermochromic Glass demonstrating its colour-changing properties through body heat. Image courtesy Inventables.

CHANNEL GLASS SYSTEMS

PROPERTIES

Cost and energy efficient

Self-supporting

Sound reductive

Length extends up to 7 metres or 23 feet, depending on windload and project requirements

APPLICATIONS

Architecture

Interior design

INFO

www.tgpamerica.com

www.pilkington.com

As glass has some structural strength, it can shaped into standardised products with the dual function of structural integrity and translucency. Pilkington Profilit ™ is a translucent linear Channel Glass System that can be arranged horizontally or vertically, with varying textures and opacities, to produce large-scale glass structures.

Because it is modular it can be used for curved walls and can be scaled to create large expanses of glass. Being made of glass, it lets light in whilst forming a structural wall and simultaneously providing privacy.

VIEW-CONTROL FILM

PROPERTIES

Alters transparency through glass by viewing angle

APPLICATIONS

Manipulate the atmosphere of a workplace

Privacy

Decorative effects

Security

INFO

www.glassfilmenterprises.com

www.glazingenhancement.com

www.lustalux.co.uk

Lumisty is a specialist view-control window film used to control visibility in contexts as broad as public architecture, restaurants, offices and conference rooms. Although not a glass in itself, upon application to panes of glass it transforms their optical properties. Its light distorting abilities enable manufacturers to control what is visible, or what is hidden, from sight. The film's appearance changes by becoming clear or frosted according to the viewing angle, with three available viewing angle options. When Lumisty goes from frosty to clear, it becomes virtually invisible.

LUMINOUS GLASS

Luminous Glass is made by silk-screening or spraying luminous pigments onto the back of standard glass. These phosphorescent pigments are activated by ultraviolet light during the day, and then emit a perceptible but unimposing glow as ambient light is reduced.

The glass is clear, hardened and coloured in either green or blue. The pigments come in two basic variations: yellow that turns to green, and light blue turning to brighter blue. The pigments can be painted onto the glass in different patterns to achieve various effects. It is mainly used decoratively but its additional ability to reflect UV light could lend it to more technical applications.

PROPERTIES

Reflects UV light

Can be applied in patterns or on differing layers to create three-dimensional effects

APPLICATIONS

Architecture

Decorative

Product design

INFO

www.materia.nl

www.si-x.nl

Luminous Glass demonstrating its greenish glow when sprayed and activated with a UV light. Image courtesy Si-x.

RECYCLED GLASS

PROPERTIES

Recyclable

Weaker than the original glass

Weather resistant

APPLICATIONS

Grit blasting

Brick manufacture

Recycled glass counter-tops

Road construction

Water filtration

Ceramic sanitary ware production

INFO

www.recycledglasslighting.co.uk

www.resopal.com

Although glass is our most frequently recycled material, the glasses produced by recycling do not perfectly match the properties of the original material. The resulting glass is often tinted, sometimes less optically clear and often less strong. In Britain, much more glass is collected than can be recycled, thus much of our glass reclaimed for recycling is exported and used in the production of new glass bottles elsewhere in Europe.

However, the recycling of glass does not only mean melting it down to produce more glass. New, innovative, and sometimes unexpected uses are regularly found for used glass, such as flux agents in brick manufacture, for recycled glass counter-tops, road construction and water filtration media.

RECYCLED GLASS FOAM GRANULATE

PROPERTIES

Light

Insulating

Fire retardant

APPLICATIONS

Public or noisy buildings

Room dividers

Acoustic panels

INFO

www.resopal.com

Recycled Glass Foam Granulate is used for its great insulating properties from sound. Resopal® make an acoustic panel consisting of foamed glass pressed into form with inorganic binders and sandwiched between particleboard. The resulting material can be used for either wall or ceiling panels. It is particularly effective in open plan office spaces as an essential sound-dampener, optimising a room's acoustics.

The use of Recycled Glass Foam Granulate means that the material is non-combustible, which makes it a great material not only for the environment, but also in fulfilling fire regulations. It can be worked using standard tools.

Resopal Acoustic panel. The cross-section shows the foam structure, which is what lends it its exceptional acoustic abilities. Image courtesy Resopal.

GLASS FOAM

Glass Foam acts as both an insulative and a structural material, taking advantage of the insulating properties of foam structures and the strength of glass in compression. It can be built directly upon, and can be tooled and worked when in slab form. It is used to fill gaps, laid under floor slabs and in a range of aggregate applications. As it is made of glass, it is inert, frost proof, fire resistant and damp proof.

Foam Glass is normally made entirely from Recycled Glass in a process that does not involve costly sorting of different types of glass that traditional glass recycling entails. All glass of varying grades can simply be ground down to form a fine powder, to which a mineral activator is added before being sintered at around 900 degrees Celsius or 1652 degrees Fahrenheit. If the material is to be used as a long block of Foam Glass it would have to be annealed like any other glass, but since much Glass Foam is used as an aggregate, it can come out of the annealer at around 300 or 400 degrees Celsius or 572 and 752 degrees Fahrenheit and undergo thermal shock, causing it to spontaneously shatter.

PROPERTIES

Cost efficient

Insulating

Load-bearing

Sustainable

APPLICATIONS

Building and construction industries

INFO

www.hasopor.com

www.technopor.com

Glass foam aggregate.
Photograh McZusatz.

AEROGEL

PROPERTIES

Commercially viable

Extremely light

Friable

Super-insulation

APPLICATIONS

Architecture

Catching stardust

Insulation

INFO

www.aerogel.com

www.aerogel.org

www.nasa.gov

When it was originally made by NASA in 2002, silica Aerogel was the lightest solid on Earth, consisting of 99.8 per cent air by volume. It is made by dissolving silica in a solvent to make a gel, and then actively evaporating the solvent through a process called "supercritical drying." This leaves behind the delicate silica framework that is extremely friable, meaning it is practically impossible to mechanically shape once formed as it easily shatters to dust.

Aerogel was actually borne out of a friendly wager between two scientists to determine who could replace the liquid with air in a gel structure first. The earliest Aerogels were developed in the 1930s, but there have subsequently been a number of improvements, not least a reduction in cost of the process, meaning that Aerogels are now commercially viable materials. A useful analogy for thinking about the structure of Aerogels is that of meringues—solid forms made by extracting the liquid from an aerated gel to leave behind a solid foam or nominally an 'egg aerogel'. It is indeed possible to make Aerogels out of other materials such as alumina or carbon that specifically results in a black material.

NASA's ultra-light silica Aerogel was developed for catching Stardust and is blue for the same reason that the sky is blue—due to the index at which it scatters the light that passes through it. The ethereal un-worldly appearance of the material, coupled with its impressive statistics and poetic use make it an alluring material for artists and designers alike. Architectural panels are currently in production that utilise the extremely efficient thermally insulative properties of the material, though they currently fail to capture the wonder of its appearance for it remains hidden or distributed as a granular form. Making large pieces of Aerogel is not only expensive by extremely difficult and limited by the dimensions of the autoclave in which it is formed. A number of people are working on making the process of manufacture more affordable but the Aerogels produced lack some of the most celebrated properties of NASA's Aerogel.

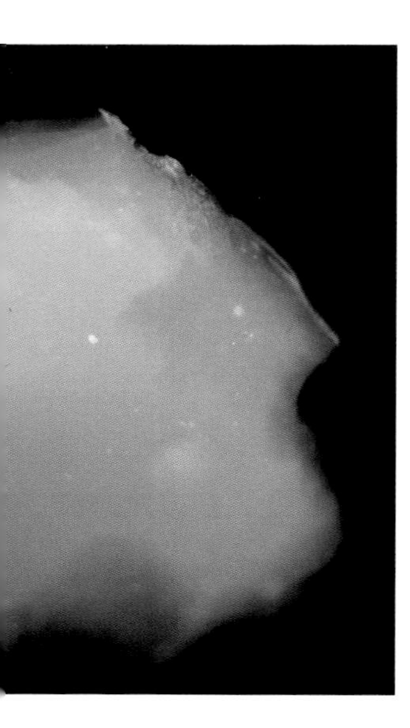

LEFT
A sample of Aerogel demonstrating its ethereal blue colour—the trait that gives it the nickname "frozen smoke".

RIGHT
Aerogel under a microscope, demonstrating it's translucency. Image © NASA.

STARDUST

Stardust, or more accurately pre-solar grains, are made from dust formed around stars that pre-existed our Sun. NASA launched a mission in 1999 under the same name on which the astronauts collected samples of the comet Wild-2 and interstellar dust by operating a 'flyby'. They returned them to Earth, or more specifically to the Jet Propulsion Laboratory at the California Institute of Technology, where the interstellar samples were examined by an international team of 200 scientists. Incredibly, these researchers found the comet particles to be abundant in organic material, those same fundamental building blocks which constitute life on Earth.

The Stardust was both captured and stored in synthetic Aerogel. Indeed, it was the development of Aerogel and its application on this mission that allowed scientists to study the Stardust at all. The mission encountered many challenges, including how to slow the particles' high speeds to capture them without harming their chemical composition, conformation, or even vaporising them completely. The device used to collect the stardust was a large metal grid filled with silica Aerogel. The speeding particles became embedded in the Aerogel, leaving 'tracks' behind them which aided the scientists in their search for particles in the structure. Aerogel was used as it is an extremely light structure and its thermally-insulating properties ensured the particles were not damaged after being embedded deep inside the spongy interior.

PROPERTIES

Organic material

INFO

stardust.jpl.nasa.gov

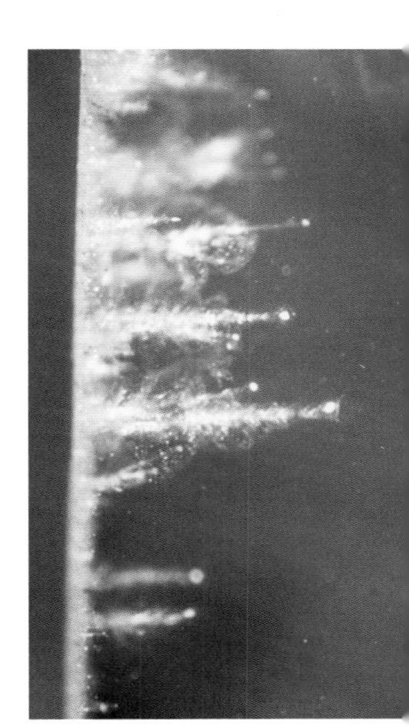

BELOW
Ennis racket-shaped dust collector. Each rectangular segment is filled with aerogel. The aerogel used by STARDUST is specially manufactured at JPL.Image © NASA

ABOVE
In an experiment using a special air gun, particles are shot into Aerogel at high velocities. Close-up of particles that have been captured in Aerogel are shown here. The particles leave a carrot-shaped trail in the Aerogel. Image © NASA.

COMMERCIALLY AVAILABLE AEROGEL

PROPERTIES

Extremely low thermal conductivity 9–12mW/mK

High porosity >90 per cent air

Nano-sized pores 20–40 nanometers

High surface area ~750m2/g

Very low tap density 30–100kg/m³

High oil absorption capacity (DBP) 540g/100g

Specific heat capacity. 7–1.15 kJ/(kg*K)

Variety of particle sizes 5 microns–4mm

Completely hydrophobic

Translucent, IR opacified and opaque

APPLICATIONS

Insulating panels

INFO

www.cabot-corp.com

Although silica Aerogel was first invented 75 years ago, Cabot has only been producing its Aerogel since 2003 at its state-of-the-art plant in Frankfurt, Germany. Cabot is the only company to develop a commercialised process that allows continuous production of the material under ambient conditions. This process allows control of porosity, pore size and distribution, and bypasses the high-cost traditional method of super-critical drying, so that Aerogel can be manufactured in a safe and continuous manner.

The Yale Sculpture Building by Keiran Timberlake Architects used Cabot Aerogel panels, making use of the material's translucency and insulating properties. The building is recognised for its energy saving credentials.

INSULATING CAPILLARY GLASS

Okalux's Kapilux capillary tube sandwich panels are made from glass tubes arranged in a honeycomb pattern. This unique structure allows Kapilux to transmit and diffuse daylight uniformly into a room, which subsequently reduces glare. Kapilux also reduces heat loss from buildings due to the cavity in each glass tube providing an insulating pocket that prevents heat transportation.

Manufactured by German firm Okalux, Kapilux is built from three panes of glass—an outer pane, middle pane and inner pane—and can be used in a variety of architectural applications, inside and out.

PROPERTIES

Can be produced with 100 per cent UV protection

High thermal insulating properties

APPLICATIONS

Architecture

INFO

www.okalux.de/en

TOP LEFT
Okalux Capillary Glass used at the The Nelson-Atkins Museum of Art, Kansas City, USA, by Steven Holl Architects.

BOTTOM LEFT
Congresszentrum Convention Centre, Zaragoza, E Nieto Sobejano Arquitectos. Photograph © Rolan Halbe, courtesy Okalux Architects.

RIGHT
A detail of an Okalux panel showing how small the fine capillaries are in comparison to the glass panels either side of them. Image courtesy Okalux.

MIRROR

PROPERTIES

Reflective

APPLICATIONS

Decorative

Microscopes

Solar power

Telescopes

Mirrors have been used in a vast variety of functional and decorative applications for many centuries. Although traditionally made of glass, some modern mirrors exploit the optical qualities of acrylic instead. The reflection arises because of the highly reflective material that is used to back the transparent pane.

Although glass mirrors have the obvious drawback of being more breakable than modern acrylic ones, they are still required in many high-tech applications such as in telescopes and microscopes, where extremely accurate engineering and minimal deformation during use are required. Technical mirrors are mostly made from Borosilicate Glass because it has a low coefficient of expansion, meaning it will not deform if it gets too hot or cold. This is hugely important when using an instrument such as a telescope to measure the precise locations over galactic distances, where even fractional deformations in the mirror can cause massive errors in the readings.

The earliest glass mirrors were backed with silver in a process that is still used by many artists today to achieve totally reflective surfaces. A silver nitrate solution is mixed with diluted ammonia to create a "silvering solution", which is then used to coat the glass, leaving a thin film of metal after evaporation. If you look at antique mirrors, you can see that this silvering process generally only thinly coated the glass and much material cracks off over the years, producing a greyed surface or flaky holes. Following the use of silver, mercury was briefly used to coat mirror backs, but for obvious reasons the health risks of this led to its demise. Most modern glass mirrors are backed with a thin film of aluminium, applied in a similar way to the silvering process, but producing a much thicker and stronger backing. Scientific grade mirrors are made with dielectric coatings such as silicon oxides and nitrides, which are applied in many thin layers. These are much more scratch-resistant than a metal coating, and are just as reflective.

Because they are wonderful reflectors, even an everyday mirror made from soda-lime glass is capable of acting as a useful tool in concentrated solar power stations. Enough energy falls on the earth every day to power all our technology for 30 years, which proves how inefficient we currently are at harnessing it. However, concentrating solar power stations are improving on photovoltaic solar generators at converting this energy to electricity, especially in areas of extreme sunlight and heat. The humble mirror

OPPOSITE
The P10 Concentrating Solar Power Tower surrounded by a field of heliostat mirrors. Image © Marco Cevat.

is revitalising the solar power industry, making solar power more efficient and cheaper to produce.

The PS10, pictured here, is a central receiver system, which means that by using a field of angled mirrors known as "heliostats" to focus the sun's heat, enough heat can be harnessed to power a steam generator, much like the ones used in power stations that burn fossil fuels.

TELESCOPE MIRRORS

THE LARGEST GLASS REFLECTORS IN THE WORLD

Innovations in polishing lenses, and casting and coating mirrors have allowed astronomers to build telescopes that can see ever further into space than ever before and consequently, further back in time. One of the main problems with extremely large glass lenses or mirrors is that, because of their size and weight, gravity causes them to sag, distorting the reflection.

The Kavali Foundation awarded their 2010 Astrophysics Prize to three people whose developments in glass and mirrors for telescopes have led to vast improvements in our ability to see into space. Their discoveries are now being employed in the latest ground telescope designs, with 16 times sharper image quality than the pictures from the Hubble Space Telescope.

Roger Angel tackled the problem of sagging by devising lighter mirrors after a series of experiments with Borosilicate Glass. He cast mirrors into a mould with hexagonal pillars, creating a pattern of holes, meaning that the mirrors could weigh a fifth the weight of conventional mirrors. Angel then adapted the conventional glass casting system by spinning the molten glass in the mould, allowing centrifugal force to pull it to the edges, forming the

LEFT
The latest artist's rendering of the European Extremely Large Telescope. Image © CSO.

OPPOSITE
Close-up view of the novel five-mirror approach of the 40-metre-class European Extremely Large Telescope (E-ELT) in its enclosure, currently being planned by ESO (artist's impression). Image © ESO.

necessary convex shape and surface of the mirror. He eventually perfected the mirror surfaces using his own computer controlled polishing device. In 2008 he cast a doughnut shaped 8.4 metre or 27.56 foot diameter mirror for the Large Synoptic Survey Telescope.

Ray Wilson took an opposite approach, making thin, flexible mirrors whose shapes were managed by a computer-controlled system attached to the back of the mirror and supported with a frame. He used a specially developed polishing tool to precisely polish them depending on how gravity would affect them. This approach is called "active optics" and was tested on the New Technology Telescope completed in 1989 and now included in European Southern Observatory's Very Large Telescope.

Jerry Nelson abandoned the idea of a single mirror and devised a huge mirror made up from smaller hexagonal tiles joined together to form a single reflective surface. All of the off-centre tiles therefore have an aspherical shape. Nelson used the natural flex of the glass to weight the mirrors, forming them all back into a spherical shape to be polished by machine. Once polished the weights are released the glass returns to its aspherical shape. The honeycomb grid of mirrors are controlled by individual computerised actuators, permanently maintaining a perfect reflecting surface.

The largest optical/near-infrared telescope in the world is planned for completion in 2022, and is called the European Extremely Large Telescope. It will be made up of five mirror stations, with the main mirror employing Nelson's formation and including almost 1000 mirrors to make a reflective surface almost 40 metres or 131 feet in diameter. Using this, astronomers predict that we will be able to gather ever more detailed information on planets and stars, the nature of the universe's dark sector, the first galaxies in the universe and super-massive black holes.

TOP LEFT
A model of the eventual configuration shows the layout of the hexagonal mirrors that will make up the largest ever glass reflector.

TOP RIGHT
The new mirror segments are being tested on their micro-controlled stands.

BOTTOM
One segment of the giant primary mirror of the E-ELT undergoing testing at ESO's facility in Germany in 2012. Photographs © ESO.

HOT MIRROR

PROPERTIES

Allows transmission of over 90 per cent of visible light

Dielectric

Greater than 95 per cent reflection of IR wavelengths

Semi-transparent

APPLICATIONS

Early digital cameras

New incandescent bulbs

Projection and illumination systems

INFO

www.edmundoptics.com

www.jnsglass.com

Although we are used to mirrors that reflect nearly all of the light that hits them, certain mirrors can be tuned in such a way as to let through only some of the incident light.

Hot Mirrors consist of Borosilicate Glass coated with a semi-transparent polarised film. The coating allows the transmission of a degree of visible light (usually around 80 per cent of the light we would normally see), whilst reflect ing over 90 per cent of the near-infrared and infrared rays. This means that such mirrors are transparent to the naked eye, but actually reflect a large proportion of the light-energy that falls upon them.

Traditionally the preserve of scientific optical equipment, Hot Mirrors have been used as filters for camera lenses and are now being developed as a coating for light bulbs and projection equipment to increase energy efficiency.

ANTIQUED MIRROR

PROPERTIES

Deliberately aged

Textured reflective surface

APPLICATIONS

Flat glass industries

Interior design

INFO

www.antiquemirror.it

The look of ageing mirror has a specific aesthetic that has seen an increase in the price of old mirrors over the years. Antique Mirror, a Tuscan company, produce deliberately aged mirrors. The look creates a texture and depth on the reflective surface akin to that of genuinely old glass, but at a fraction of the price.

In order to achieve the effect, the silvered backing is oxidised in random places, mimicking the effect of age on the mirror back.

The patina of 'antiqued mirror' where flecks of oxidation appear. Image courtesy Antique Mirror.

REFLECTIVE TAPE

Light can undergo very strong reflections at the internal surfaces of glass forms, a phenomenon which is harnessed in Optic Fibres. However, such a mechanism can also be used to create reflective safety strips that reflect a car's bright headlights back at the driver.

Scotchlite fabric by 3M is coated with a layer of very densely packed tiny glass beads embedded in a reflective silver coating. This creates, in effect, a large array of microscopic concave mirrors that reflect the light back in the direction from which it came.

The process that produces these glass beads involves blasting molten glass through nozzles with jets of air. This creates a glass spray, with each individual droplet solidifying as an individual bead. During the spraying the beads become coated with titanium tetrachloride and wax, making them very shiny and practically scratch resistant.

Metlon, a decorative fabrics company in Rhode Island, USA, make the tape into strips, like a yarn, which can be woven into garments, used as decoration, or worked into ordinary clothing.

PROPERTIES

Highly reflective

Can be made into adhesive tape

Available in strips for safety clothing

APPLICATIONS

High visibility garments

Sculpture

INFO

www.metlon.com

solutions.3m.com

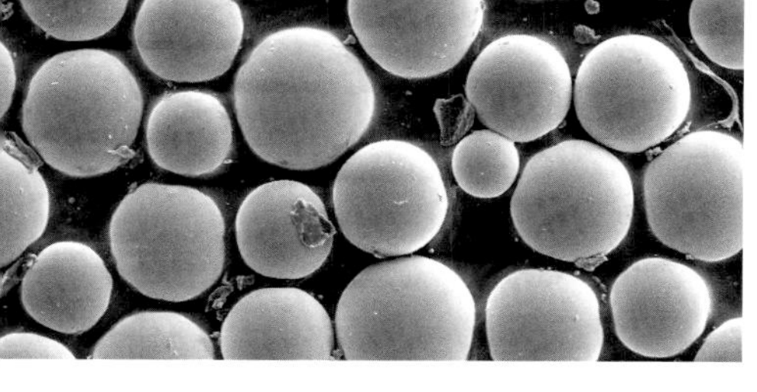

LEFT AND BELOW
Microscope images of the tiny glass beads embedded in the fabric, the cause of the retro-reflection.

RIGHT
Reflective Tapes being activated by light. The retro-reflection makes the tape appear to emit light.

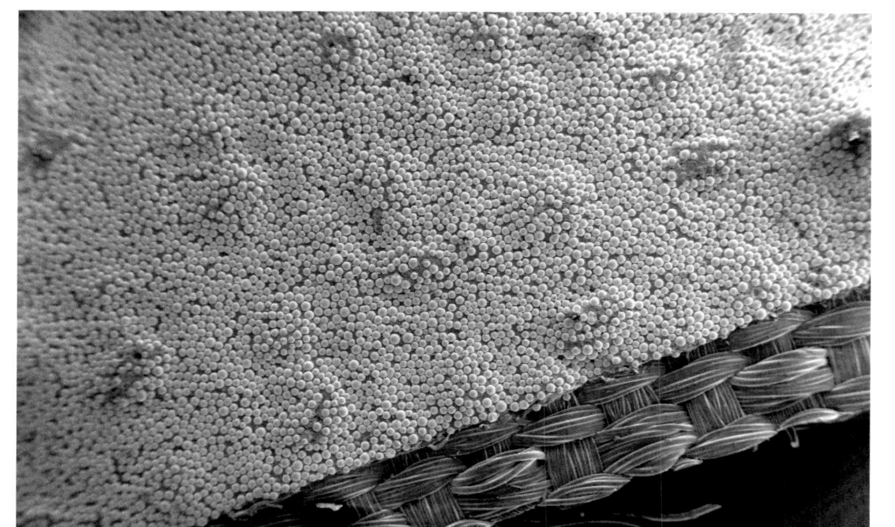

DECORATIVE GLASS SPHERES

PROPERTIES

Fire resistant

UV resistant

Interacts with light

Can be used on structural surfaces

APPLICATIONS

Interior design

Furniture

INFO

www.si-x.nl

Globo-X is a textured material made from a 'carrier' structure (such as steel, aluminium or glass) that is decorated with spherical glass elements available in a variety of shapes, sizes, patterns and even colours.

Mainly used for aesthetic purposes, Globo-X can be used inside and out, as a wall covering, on ceilings, as a lighting element and more.

Embedding glass spheres in a flexible polymer creates a fabric-like material with interesting visual properties, which can be used with great flexibility in wet or dry conditions. Images courtesy Si-x.

LASER ETCHED GLASS

Made using a laser that engraves beneath the surface of the glass, 3-D Engraved Glass is etched with an image or pattern to produce a three-dimensional design.

Available in a selection of colours and thicknesses—from 8 to 25 millimetres or 0.3 to one inch—3-D Engraved Glass is suitable for a range of applications, mainly decorative. It is particularly effective when used with back lighting.

PROPERTIES

Low maintenance

UV resistant

Weather resistant

APPLICATIONS

Architecture

Decorative

INFO

www.lasercrystal.co.uk

www.si-x.nl

TOP
Detail illustrating the way in which the laser 'damages' internal areas of the glass, thereby etching the interior with a subtle design. The fracture caused by the laser interferes with the optical qualities of the glass, which is what makes the tiny fracture lines visible. Images courtesy Si-x.

BOTTOM
The process, above, can be scaled up, creating large visual patterns for architectural forms. Images courtesy Si-x.

OPTIC FIBRE

PROPERTIES
Flexible
Optically pure

APPLICATIONS
Scientific
Telecommunications

INFO
www.fibercore.com
www.fibre-options.com
www.horiba.com

Optic Fibres for telecommunications are made from extremely pure optical glass which is extruded into long threads. A glass blank is made from which the fibres are pulled. As the blank is perfectly round and the thread is pulled directly downwards, the cross-section of each fibre remains perfectly rounded. The long fibres are then wound onto a spool like a thread. The glass is made from a pure mixture of silicon and germanium, which is reacted with oxygen to make their respective oxides, and formed through a modified chemical vapour deposition (MCVD) procedure.

Although in essence the glass is similar to soda-lime or Borosilicate Glasses, it is the extreme purity that sets it apart. The MCVD process is performed with exceptionally pure ingredients under sterile conditions, so that the impurities that are inevitable in normal glasses cannot find their way into these optical glasses.

The fibres are checked for tensile strength, reflective index, attenuation and bandwidth. It is claimed that, such is the optical purity of this glass, if it filled the ocean you would be able to see through it to the sea bed, even at the ocean's deepest point.

LEFT
Optic Fibres demonstrating the way in which light is carried through the length.

RIGHT
The actual glass of the cable is only a tiny fraction of the cable's diameter. To protect the fragile cable it is surrounded in numerous protective layers. Image courtesy Global Marine Cable.

INTERNET CABLES

With the number of wireless devices in use at the moment it is tempting to believe that cables are a thing of the past, but more internet cables are being laid on ocean floors than ever before. What is more, despite their diminutive size (most of them are a mere seven centimetres or 2.76 inches in diameter) each one can theoretically host up to 50 million phone calls at any one time.

Undersea internet cables use the most advanced Optic Fibre technology and the purest optical glass possible. Each cable is made up of a core, which is a thin glass strand, a cladding, which is the outer optical material, and a buffer coating, which is flexible but protective. There are two different types of fibres; single-mode fibres, which transmit laser light and have much thinner cores, and multi-node fibres which transmit infrared light and have larger cores.

Vast underwater cables have to carry coded information over huge distances. As with any code, there has to be a process of coding and decoding to make sensible use of such signalling. Firstly, a transmitter encodes digital information into light pulses that are then channelled down the optical fibre. A long way away but only a split-second later (as light travels at 300,000 kilometres or 186.4 miles per second) an optical regenerator recharges the light with a laser, boosting it ready for decoding. Finally, an optical receiver processes the signal and decodes it into electrical signals to be fed into devices such as televisions, telephones and computers.

PROPERTIES

Flexible

Minuscule

Protective coating

Transmit light

APPLICATIONS

Underwater network for internet connection

Coding information

INFO

www.batt.co.uk

www.globalmarinesystems.com

TOP LEFT
Even the longest optical cables could not traverse the whole Atlantic ocean. Cables are therefore joined via these cable connectors.

BOTTOM LEFT
A cable is laid off the back of a boat and embedded in the ocean floor.

RIGHT
A spool of Internet Cable waiting to be laid. The scale of the spool demonstrates the vast distances the cables must cover. Images courtesy Global Marine Cable.

GLASS WOOL

PROPERTIES

Acoustic and thermal insulator

Cost-effective

Non-combustible

Non-toxic

Skin irritant

Very low moisture absorption

APPLICATIONS

Architecture

INFO

www.glass-wool.com

www.polyglass.com.my

Glass wool is typically made from a combination of recycled glass and raw sandy materials that are cleaned and then melted. The molten mixture is fed into a centrifuge with perforated walls, which when spun, extrudes the glass through the holes into thin stringy fibres which cool immediately, creating tiny cotton-like threads of glass.

Although this process has been used since the 1930s, it is subject to continual innovation. For example, combining glass fibres with new resins and backing materials enhances specific structural properties of this material. It is an extremely good acoustic and thermal insulator, comparatively cost effective, non-combustible, non-toxic and has very low moisture absorption.

Glass Wool, when combined with certain resins, forms Fibreglass, a hugely important and pervasive composite material. To make rolls of Fibreglass insulation, masses of threads are fed along a conveyer and sprayed with resin, which helps to bind them. A backing is often added which is either a resin-soaked paper or thin aluminium, both of which conform to strict fire regulation codes.

Despite being technically non-toxic, glass wool is an irritant and harmful to the body. Without the binder holding it together it is particularly bad for the lungs as tiny shards can be inhaled or lodged in the skin.

3-D GLASS FABRIC

PROPERTIES

Weight reduction as compared to metals

Fire-resistant

APPLICATIONS

Flooring

Seating

Air ducts

INFO

www.vetrotex.com

bola-triglass.nl

Bola TriGlass is a Spacer Fabric made entirely of glass. Used mainly as a core material to be embedded with thermoset resins, it can be adjusted for a variety of applications. It is made in the same process as for Spacer Fabrics, and can be used as both a surface and a reinforcing material.

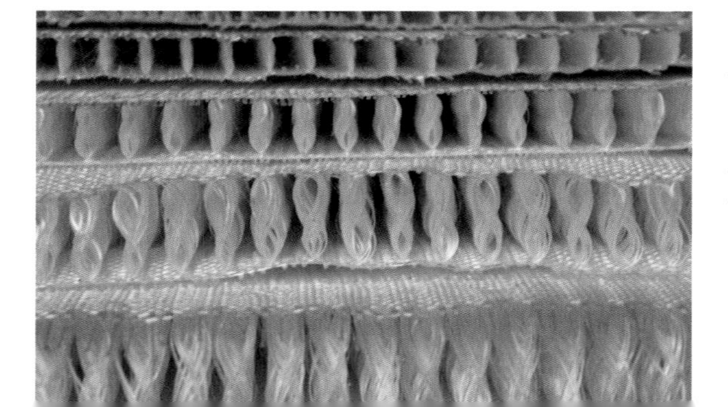

Cross-sectional samples of glass Spacer Fabrics. The stiffness of the vertical distance yarns creates bridges between the woven layers. It is this stiffness that controls the give of the material when compressed. Varying the weave in both directions therefore affects the material's properties. Images courtesy Bola Triglass.

MICROSCOPIC GLASS FLAKES

Microscopic Glass Flakes are used in industrial applications where they are mixed with other materials as a component for composites. For example, they can be added to paints to significantly enhance corrosion resistance. The Glass Flakes, made from C-type Borosilicate Glass, are able to orientate themselves during painting so that they form a protective barrier against a potentially harsh chemical environment.

Although this paint is technically a Composite, it is worth mentioning here as it is the innate characteristics of the glass that gives the paint its anti-corrosive properties, and make it stable even in marine applications, where unprotected metals are at risk of the powerful corrosive action of salt water.

PROPERTIES

Microscopically small

Corrosion resistant

APPLICATIONS

Paints

Composites

INFO

www.kinera.no

LUSTROUS GLASS PAINT

Miraval, developed by Merck Chemicals, is a pigment comprised of Microscopic Borosilicate Glass Flakes coated in titanium dioxide. The inclusion of the flakes within glass creates a glittery lustre, reflecting large portions of the colour spectrum in an iridescent display. The pigment achieves its lustre by mimicking the effect of a pearl, which is comprised of alternating layers of proteins with high and low refractive indices. The pigment is constructed with microscale control, using an extremely thin layer of metal oxide on the glass. Iridescence arises due to the way light interacts with these layers.

Lustrous glass pigments can also be combined with each other to create specialist surface finishes and effects. As the layers are exceptionally thin, they can be overlaid without obscuring those underneath, creating additional effects through the combination of finishes.

PROPERTIES

High brilliance and gloss

Iridescent display

Outstanding sparkle

Unique rainbow effect

As small as ten micrometre particle size

APPLICATIONS

Specialist surfaces and finishes

INFO

www.merck-chemicals.com

Tiny Glass Flakes create a glittery inclusion in paint. Test sample demonstrating the extreme lustre achieved on a curved surface. Image courtesy Merck Chemicals.

CAST GLASS

PROPERTIES

Can be clear or coloured

Variable forms can be created

Transmits light

Reflects light

Can be cut and polished

APPLICATIONS

Architecture

Sculpture

INFO

www.jcdainc.com

www.tgpamerica.com

Although casting glass is an old process, it is still used in innovative ways. Strong, structurally interesting and beautiful, with an opacity ranging from completely translucent to totally opaque, Cast Glass has been used for glass bricks, panelling, flooring, steps and roofs for some time.

Another architectural innovation in glass casting can be found in interior applications where there is little light. Veluna luminescent glass, for example, includes phosphorescent agents that allow it to glow in the dark. Current applications include use in pathways, but it is mainly used as a decorative feature in interior spaces.

Glass casting is also the process of choice for many sculptors working with the material. Adding certain mineral compounds colours the transparent glass, and allows control over the amount of light it transmits, whilst the casting process enables the creation of specific predetermined shapes of relative precision.

Casting Glass allows a designer to manufacture specific shapes for a particular function. James Carpenter Design Associates used the process for *Ice Falls* as illustrated below.

For *Ice Falls*, Hearst Tower, New York, James Carpenter Design Associates created a water cascading area with specially developed Cast Glass crystals made by pouring molten glass into a graphite mould and annealing the forms. The practice developed the optimal size and shape of the Cast Glass to create the effect of the internal reflections of cut diamond.

ABOVE
The casting process. Each individual segment was poured and annealed this way.

RIGHT
View of the Cast Glass with water flowing. The accent blocks emerge above the water to provide a mirror-like reflection. Cool water running over the glass acts as a filtration and cooling system for the building. Images courtesy James Carpenter Design Associates

CRYSTAL GLASS

Crystal Glass is not actually crystalline at all, for, like all glass, it has an amorphous microstructure. Invented in the seventeenth century by George Ravenscroft, Crystal Glass was given its name because it looked like rock crystal and produced a similar visual effect.

For a glass to be called Crystal Glass nowadays, it has to contain more than 24 per cent lead oxide. The most famous examples of lead crystal are Swarovski crystals, which contain 32 per cent lead oxide, utilising a secret family glass recipe known for its superior clarity and sparkle.

Lead crystal has different properties to the soda-lime glasses more commonly used for windows and low-cost glassware. Because lead is a heavy metal with a high atomic number, it produces a glass that it is denser than soda-lime glass and noticeably heavier in the hand. It has a high refractive index that makes the glass particularly sparkly, and therefore the material of choice for chandeliers and prisms.

Lead is highly poisonous and therefore those working with it must take specific precautions. The Canadian government even recommends specific washing instructions for lead crystal glassware, for example rinsing it in vinegar to prevent metal leaching.

Crystal Glass does occasionally refer simply to cut glass, and although the only cut glass with the official name "crystal" is made with lead oxide, most prism-like cuts get called crystals because of their patterns of refraction.

What sets Swarovski crystals apart is the precision of form and quality of sparkle—which in part comes from their high lead content and in part from the method of grinding. Swarovski created a machine capable of precision grinding glass in the same way that diamonds are ground. The most intricate crystals have as many as 100 cut and polished facets.

PROPERTIES

Sparkle

High internal reflection

Interesting optic effects

APPLICATIONS

Decoration

Jewelry

Design

INFO

www.swarovski.com

The Centenar, at 300,000 carats, is the world's largest cut crystal. It is cut with 100 facets, representing the 100 years that Swarovski had been in operation in 1995. Image courtesy Swarovski

DESIGNING WITH CRYSTAL

SWAROVSKI CRYSTAL PALACE

Swarovski Crystal Palace is an experimental platform that celebrates the possibilities of lighting using Swarovski crystals when teamed with renowned artists and designers. Cut crystal is crafted into innovative presentations, showing the chandelier in a new light. In a relatively recent branch outwards from the fashion and jewellery worlds in which Swarovski is grounded, Crystal Palace is focuses on the power of lighting.

The output of New York architecture studio Diller Scofidio + Renfro combines architecture, design and theatre, embracing all varieties of commissions, and for institutions such as MOMA, New York, and the Museum of Contemporary Art Los Angeles. Their *Light Sock* overturns the typical arrangement of crystals in a chandelier, instead being made up of netting filled with Swarovski crystals, reminiscent of a buoy wrapped with rope mesh. A light bulb in the centre of the crystal mass increases the glow outward from the centre of the elegant design.

Arne Quinze is the founder of design agency Quinze & Milan, a springboard for the efforts of like minded designers. For *Dream Saver*, Quinze used five and a half kilos of Swarovski crystal, hanging down like branches of a delicate willow tree, to create a bower or tunnel structure, emphasising the life energy he wished the project to emit. The skeleton of the piece is made of metal and wood, 12 metres or 40 feet in length and three metres or 10 feet in height, covered in fibreglass sheets. Tapping into the viewer's emotional response, Quinze gives the crystals a life of their own by the body of work moving around the participants, their reactions haunting the installation, gently swaying and glinting, enhancing the experience for the next participants.

RIGHT
Light Sock by Diller Scofidio + Renfro for Swarovski Crystal Palace. Image courtesy Swarovski Crystal Palace, photograph © Ken Hayden.

OPPOSITE
Dream Saver by Arne Quinze for Swarovski Crystal Palace, Image courtesy Swarovski Crystal Palace, photograph © Leo Torri.

BIOGLASS

PROPERTIES

Can be carved, tooled and formed easily

Can be made into a Composite material with human bone for reconstructive purposes

Powder or solid

High bioactivity

Not load bearing

APPLICATIONS

Dental

Medical—repairing bones and grafting tissue onto bone

Microchips designed to stay in the body

INFO

ufdc.ufl.edu

Bioactive glasses, commonly referred to as "bioglasses", are unassuming materials with amazing capabilities. Made from SiO2, Na2O, CaO and P2O5, bioglasses have a much lower content of silica than traditional soda-lime glass. This means that some varieties of Bioglass can be completely absorbed into the body, and are thus used in bone reconstruction surgery and to graft tissue onto bone. The Bioglass is "seeded" with a patient's stem cells and then positioned in the body in place of missing bone. The native bone then grows through the Bioglass, using it somewhat like a scaffold, consuming the Bioglass as it goes and leaving behind a seamless repair.

Bioglass is the kind of material that reminds us that our bodies consist of the same elemental matter as everything around us, animate or inanimate. It is this concept that is leading to ever more advanced 'biomaterials' that can repair and enhance our bodies.

Early Bioglass was developed by Professor Larry Hench at the University of Florida in the 1960s. He was trying to develop ways to regenerate human bone as a response to the urgent needs of the many wounded soldiers returning home from combat in the Vietnam War.

Looking to the future, it will be possible to use bone scanning technologies and the 3-D printing of Bioglass to create highly accurate bespoke bone replicas—a great example of where material and process innovations can come together to create vast leaps in technological capability.

Bioglass can be engineered into the forms required for implantation. As such they are powdery and aerated structures which are easy to shape and allow damaged bone to grow through.

VITRIFIED GLASS

A sample of nuclear waste converted to glass through Kurion's process. Image courtesy Kurion.

PROPERTIES

Visually similar to obsidian

APPLICATIONS

Contamination reduction

INFO

www.kurion.com

Vitrification is the process of turning something into a glass, and is usually achieved by the rapid cooling and solidification of a liquid substance.

An example of its common usage is as a way to deal with toxic waste. In this process, liquid waste has some glass-forming additives mixed in, and then it is rapidly cooled. The resultant solid is much easier to handle than the original waste, eliminating many concerns over air or groundwater contamination.

Historically, vitrification has involved transporting dangerous waste to sites where the process can be performed. However, Kurion, a specialist company in this area, have recently begun to revolutionise vitrification for they are able to take the process to the waste by inventing modular vitrification systems (MVS). This substantially reduces difficulties associated with transportation and makes disposal safer and easier. The vitrified waste looks like obsidian, the naturally occurring volcanic glass.

LIQUID GLASS

Liquid Glass, "water glass" or more correctly sodium silicate solution, is a material with the potential to revolutionise how we protect surfaces and structures, making them hydrophobic and chemically resistant on a domestic and industrial scale. An example of this is the use of Liquid Glass to plug leaks at Japan's Fukushima Nuclear Power Plant in 2011.

Soaking materials in a Liquid Glass solution, or spraying them with it, creates an ultra-thin layer of silica that lends them the properties of glass. Wood impregnated with Liquid Glass becomes glass-like on the surface, with counter-tops and door handles repelling germs, and fabrics becoming waterproof as a result of their being combined with Liquid Glass. And as it is only around 100 nanometres thick, Liquid Glass remains highly breathable and flexible.

PROPERTIES

Flexible

Highly breathable

Only around 100 nanometres thick

Repels germs

APPLICATIONS

Nuclear Power Plants

Hydrophobic and chemically resistant surfaces

INFO

mistralni.co.uk

CERAMICS

When one thinks of ceramics, images of china plates or stoneware pots often come to mind. Objects made from a material that is transformed from wet slippery clay into a hard and brittle substance that is liable to chip, crack or shatter. Despite this fragility, ceramics are in fact extremely durable materials, unmatched in their resistance to corrosion and decay. The fact that our knowledge of ancient civilisations is often based on excavated ceramic artefacts is testament to this. They are very strong under compression, can show extraordinary heat resistance, and are easily moulded before firing. However, without exception, all ceramics are brittle and despite being strong, they are definitely not tough.

The material make-up of ceramics is an interesting combination of metallic and glassy characteristics. The structure is similar to that of metals, with crystalline grains forming and fusing to build a solid mass. However, the presence of non-metals such as oxygen, carbon and nitrogen, combined with silicon and various metals, leads to a more intricate and complicated structure than in metals, with amorphous glassy phases forming amongst the crystalline grains. This complexity in chemical constituents and structure means that ceramics can be engineered to exhibit a huge range of different properties, tailored to an individual application.

Clay, an abundant and easily extracted natural resource, forms the basis of traditional ceramics. The chemical composition of clay varies depending on its geographical origin, but they are all based on silicate minerals with assorted metallic constituents. When wet, these mixtures form a mouldable mass, which is then dried and fired to create the final ceramic. During firing, any remaining water evaporates and silicate crystal grains form. A glassy silica phase also forms during firing which spreads throughout the structure, acting as a solid glue between the crystalline grains. Although this sort of structure is rigid and strong, the boundaries between the different phases provide a highway for crack propagation, giving rise to the brittle character of this material. The situation is further confounded by the presence of pores which form in almost all ceramics.

Modern engineered ceramics can exhibit some extraordinary properties such as ultrahardness and extraordinarily high heat resistance. Traditional ceramics are used to make objects such as bricks and tiles, whereas engineered ceramics are used as high-performance bearings, ultra-strong turbine blades, super-sharp knives and as heat-resistant shields for space

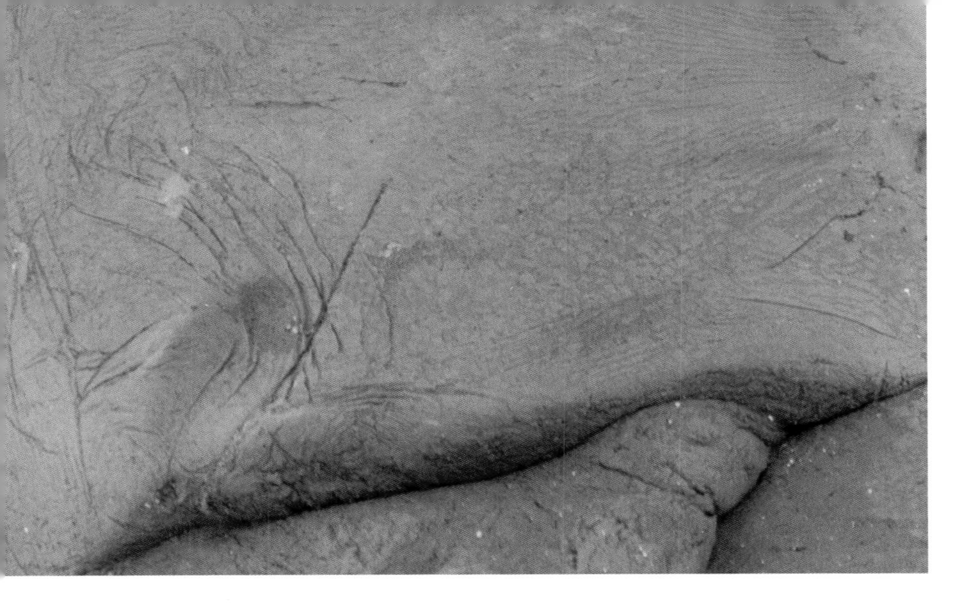

vehicles. Instead of being moulded as a wet mass, these ceramics generally start off as powdered ingredients mixed with a polymeric binder. This mixture is then shaped by methods such as extrusion, pressing, casting or injection moulding, before being fired to sinter the powder into a solid mass. A particularly impressive example is the Silicon Nitride ball bearing, which exhibits extreme hardness second only to diamond.

Ceramics have a long legacy in the history of materials, but they are not just a material of the past—their extreme durability and the advent of high-performance engineered ceramics has ensured that they are very much a material of the future as well.

PH and ZL

PREVIOUS PAGES
Detail of micher'traxler's Cauliflower bowl as part of their reversed volumes series.

TOP AND BOTTOM
Details of the contrasting textures of Terracotta and Porcelain clays. The fine grain size of Porcelain creates much smoother surfaces.

TERRACOTTA

Despite the fact that Terracotta ceramic has been around for centuries, designers are continually finding new and interesting ways to work with this material. It is an enriched mouldable clay which can be air-dried or fired in a kiln to fix it into a hard solid form. Its insulating properties, discovered around 1400 BC, have made it invaluable for use in masonry and construction, alongside kitchen and tableware.

Architectural Terracotta, in both its glazed and unglazed forms, is sturdy, relatively inexpensive, and can be moulded into richly ornamented detail, making it a material of choice for both structural and decorative forms through the ages. The brightly coloured facade of the Brandhorst Museum in Munich is a recent example of how glazed Terracotta is still used in innovative ways. This facade was manufactured by NBK who specialise in this type of modular architectural ceramic. The properties of the material and the mode of modular construction also enable the facade of buildings clad in such a way to absorb sound—a rare characteristic of buildings in contemporary urban environments.

Some of the world's most advanced water treatment technologies involve the use of Terracotta-based products. Ceramic water filtration projects have improved drinking water quality where their use has been implemented.

PROPERTIES

Sonorous

High load-bearing

Fireproof

Commonly associated with deterioration problems

APPLICATIONS

Pottery

Architecture

Water filtration

INFO

www.nbk.de

OPPOSITE
The facade of the Brandhorst Museum in Munich demonstrates the bright glazes possible in the range of architectural products from NBK.

LEFT
Details of the Brandhorst Museum wall illustrate the formation of the wall structure, where the batons form a screen with space between the outside wall and the screen. This is what helps to insulate the building from sound and also adds depth and texture to its facade.

RIGHT
The texture of Terracotta is relatively rough compared to other Ceramics. Its larger particles lend the glaze on the NBK batons its interesting texture.

A sample of Kaolin. This block is actually only three centimetres in width. The magnification demonstrates how fine and powdery the material is.

KAOLIN

PROPERTIES

Smooth

Fine

Rheological

APPLICATIONS

Plastics

Commercial paint

Drugs

Paper coating

Kaolin is a soft, white, mineral-rich clay that is widely mined the world over. As clay, it is the main constituent of Porcelain and has a long history of use in the potteries of ancient China, where it was prized for the brilliant white lustre that it afforded. Today it is commonly used as a filler in plastics, paper, paints and pills, finding its way into many everyday household items. For example, particles of Kaolin are to be found in many types of toothpaste where fine grains of the material are added to clean and polish the surface enamel of teeth. Being chemically inert over a wide pH range, it is even occasionally added to food as an anti-creaking agent, more commonly listed as E559.

If one zooms in to look at the compositional structure of Kaolin, one finds it is composed of fine-grained plates of aluminium silicate that produce specific optical properties. As a result, it is commonly used in the paper industry as a coating that enhances the tonal quality and colour of white paper and helps to provide a smooth glossy finish. Kaolin is also used as a pigment in its own right and is frequently to be found in household paints as an accompaniment to and extender for expensive white pigments like titanium dioxide. Its slight rheological properties are also useful in liquids like paint or molten plastics for this helps to keep the pigment evenly dispersed throughout the liquid.

PORCELAIN

Porcelain was originally discovered in China around 600AD —the earliest Porcelain, commonly called "primitive Porcelain", appeared during the Shang Dynasty. Porcelain is classically delicate and graceful, ghostly translucent and surprisingly tough. Its translucency and strength arise mainly from the formation of glass and the mineral mullite within the body of the material when fired at temperatures of around 1200 degrees Celsius or 2192 degrees Fahrenheit.

Historically, Porcelain is associated with the manufacture of ornate figurines and high-quality tableware. Today, designers and makers still work with the material because of its toughness, smooth texture, visual qualities and cultural associations. In the *Anamorphosis Milk Bottle* pictured, designers Sofie Lachaert and Luc d'Hanis have combined a traditional application of Porcelain with an anamorphic image, taking advantage of the fine, smooth surface that can be achieved with the material, and coating it with a reflective glaze, mimicking a mirror.

Porcelain is also widely used to enhance teeth in cosmetic dentistry in the form of veneers. Comparable to tooth-enamel in strength, Porcelain dental veneers are ultra-thin, custom-made laminates that are bonded directly to the teeth with resin adhesives.

PROPERTIES

Smooth

Hard

APPLICATIONS

Dentistry

Tableware

Lighting and product design

INFO

www.droog.nl

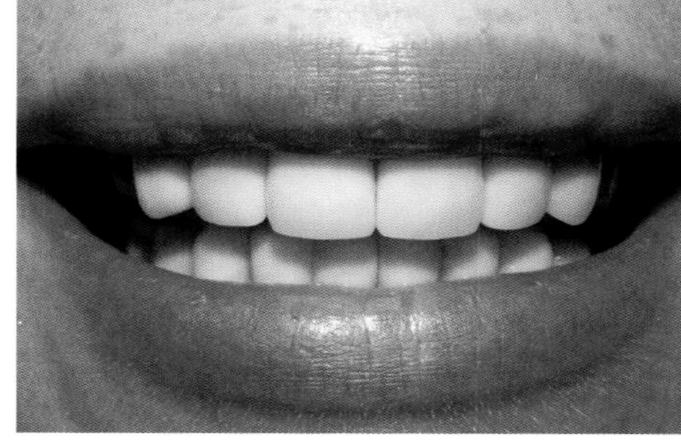

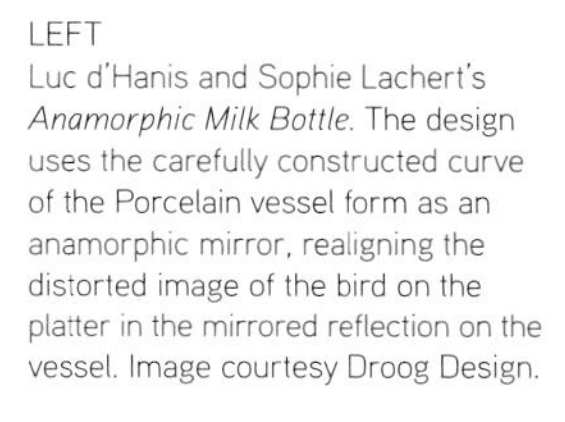

LEFT
Luc d'Hanis and Sophie Lachert's *Anamorphic Milk Bottle*. The design uses the carefully constructed curve of the Porcelain vessel form as an anamorphic mirror, realigning the distorted image of the bird on the platter in the mirrored reflection on the vessel. Image courtesy Droog Design.

ABOVE
Porcelain veneers use in dentistry.

MOULDED TILES

PROPERTIES

Scratch-resistant

Durable surface

APPLICATIONS

Interior and exterior design

INFO

touchy-feely.net

Ceramics are the perfect material for creating smooth, hygienic and tactile surfaces due to the many ways they can be moulded and formed and the variety of different finished that can be achieved with them.

A novel use in tile form is Found Space Tiles. These are ceramic wall tiles that are intended to encourage tactile engagement with their surface. The tiles are intended to form clusters with their raised, undulating surfaces referencing the human body.

Produced by Dutch company CorUnum, Found Space Tiles are made from earthenware ceramic, which is slip-cast using a plaster mould to create a hollow shell, six millimeters in thickness.

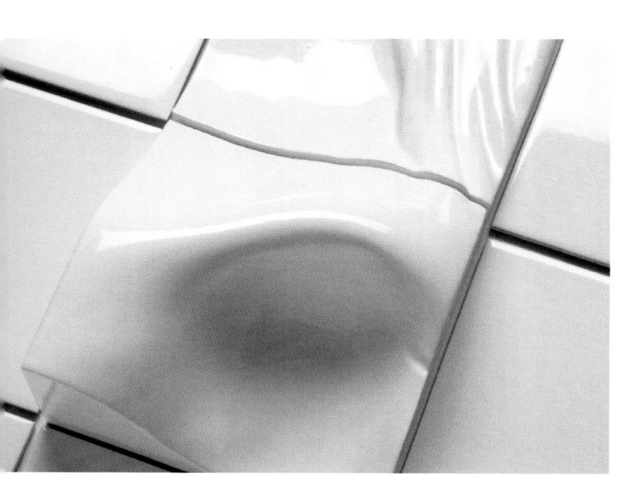

These tiles can be used to make use of the reflection from the glaze's glossy surface. By creating raised undulations the tiles catch the light, creating a silky, undulating wall formation. Images courtesy Touchy Feely Haptic Design.

ZIRCONIUM EMBEDDED TILES

These tiles can be produced at a minimum thickness of three millimetres.

Kerlite is a ceramic material with zirconium grit embedded within it. It is made from a standard selection of clays, Kaolin and feldspars, but is pressed and fired using an innovative technology.

The inclusion of zirconium enables the manufacturers to make the sheets incredibly thin, thus much lighter in weight than regular ceramic tiles. It is resistant to fire, does not stain and can be used for internal and external applications.

PROPERTIES

Available in a range of sizes

Six colour variations

Resistant to dirt

Light

Fire-retardant

APPLICATIONS

Architecture

INFO

www.cottodeste.it

MICROWAVABLE CERAMICS

Pure ceramic dishes are labelled as "microwave safe" as they do not interact with microwave radiation, and therefore do not heat up directly from microwave irradiation. However, the food which is heated loses heat to the ceramic, which actually reduces the efficiency of the heating process.

Sridhar Komarneni, a Professor of Clay Mineralogy at Penn State, working in conjunction with research laboratories in Japan, has developed a ceramic that can be heated in the microwave. Made with small amounts of iron oxide, the Microwavable Ceramic is an iron oxide-petalite foam. The material works in two ways; the iron oxide responds to the microwave radiation and heats up, and the petalite acts as an insulator, helping to retain the heat.

The material has potential for use in science, perhaps more than domestic uses, because microwaves can be used to decontaminate experiments from organic contaminants such as oils.

PROPERTIES

Heats to 278 degrees Celsius or 532 degrees Fahrenheit

Microwavable

APPLICATIONS

Scientific experimentation

Potential for cookware

INFO

cropsoil.psu.edu

NON-FIRE CERAMICS

PROPERTIES

Safe

Non-toxic

Oil based clays are reusable

Fine texture

APPLICATIONS

Model-making

Casting

Toys

INFO

www.amaco.com

www.michertraxler.com

There are a number of different types of clay that do not require firing called "modelling clays", most of which can be found in craft shops. A majority are dispersions of minerals in suspending agents, which behave much like firing clays that are in their modelling stages, but are not as resistant and tough when dry. Polymer clays such as Sculpey or Fimo are classed as modelling clays but do not contain clay minerals, so although they are often called "clays", they are not considered ceramics.

Oil-based clays are mineral clays suspended and bound with oil, which prevents the clay from drying out. Traditionally used for model-making for industrial design and animation the clays, such as Plasticine, can be used to temporarily make watertight seals for casting, as well as making the model to be cast. As it does not dry out, the material can be constantly re-used, making it an extremely versatile tool for artists and designers.

A form of air-dry clay known as "wet porcelain", which is useful for sculpting or taking impressions, actually contains no ceramic at all. It is mainly used for making model flowers with delicate and porcelain-like forms, hence its name. It could actually be made from materials found at home—cornstarch, PVA, mineral oil and lemon juice, mixed to produce a smooth paste. This paste can then be coloured using ordinary acrylic paints, and, unlike most air-dry clays, the mixture barely shrinks when dry, making it perfect for moulding objects and taking very fine impressions. In order to use the 'clay' as one might use traditionally fired ceramics it must be sealed against moisture, using a sealant such as shellac.

Design duo mischer'traxler used a Non-fire Ceramic to create a series called 'reversed volumes'. The design uses ceramic powder, which becomes very hard without being fired, meaning that each piece is as unique as the fruit or vegetable used. 'reversed volumes' was initially developed for FoodMarketo, a pop up shop organised by DesignMarketo, *Apartamento Magazine* and Marion Friedmann during Milan Design Week 2010. Dimension: from ten–22 centimetres.

ABOVE
A selection of the finished forms with the vegetables and fruits that they were moulded from. The design celebrates the soft colours of the fruits and their individual forms and textures.

LEFT
A section of testers for the process of colouring the bowls. Using pigment to dye the Ceramic creates resultant forms that resemble the fruit or vegetable from which they were moulded. Images courtesy mischer'traxler.

ALUMINA ADVANCED CERAMIC

PROPERTIES

Extremely hard

Good electric insulator

Amphoteric

Soft translucent surface

Insoluble in water

APPLICATIONS

Used for cutting tools

INFO

www.ceradyne.com

www.hydro.com

You would not necessarily think a ceramic could make great body armour. Most of us have dropped a cup and watched it smash from only a relatively small impact. However, it is the very fact that ceramics are brittle—and therefore shatter—which makes this Alumina Ceramic work as body armour. By shattering on impact, the Alumina disperses the impact energy of a bullet through the material, thus stopping it from carrying on into the body and harming the wearer.

Alumina body armour is much lighter in weight than previous armours, which are traditionally made from metals. The ceramic plate is contained within an extremely strong Polymer mesh such as Kevlar, which holds the shattered ceramic together after impact.

As a material, Alumina is used in many varying applications. However, the vast majority is consumed by the manufacture of aluminium in the Hall-Heroult process. Here, the elemental aluminium is extracted from the Alumina by soaking it in a hot salt bath. Alumina is also useful for electrical insulation due to its very low electrical conductivity but relatively high thermal conductivity.

LEFT
A pile of Alumina in its processed form. Image © Norsk Hydro.

BELOW
A bullet-proof vest contains panels made from advanced ceramic. The ceramic makes the vest light while remaining strong. Photograph PEO Soldier.

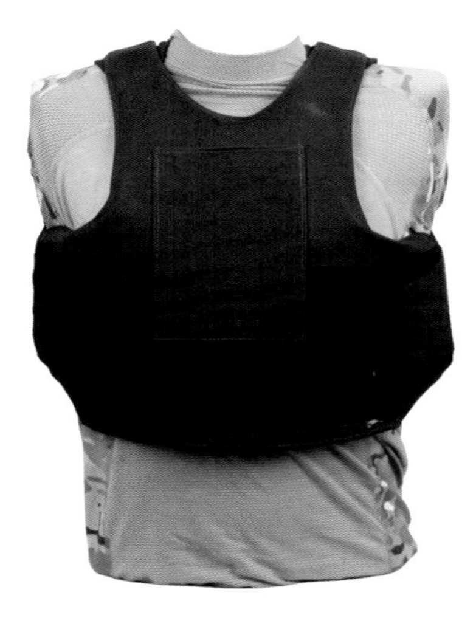

SILICON NITRIDE

Silicon Nitride, a compound of silicon and nitrogen, is a super-hard Ceramic. It exhibits exceptional strength over a broad temperature range, and its ability to withstand the extreme environment inside a jet engine is why it is the material of choice for NASA's Space Shuttle bearings.

Despite being relatively light and feeling unimposing in the hand, if a Silicon Nitride ball bearing is dropped on a floor it will leave a dent, even in concrete, as the ball does not deform at all upon impact so all the energy is transferred straight into the ground. In fact, the only materials harder than Silicon Nitride are diamond and Boron Nitride.

Developed by engineers in the 1960s, Silicon Nitride has since been used in various high-tech applications such as aerospace and military equipment. However, it is only now beginning to find its way into more commercial applications such as diesel engines, insulators in integrated circuits, bearings in saltwater fishing reels and even in-line skates.

PROPERTIES

Incredibly hard and smooth

Low coefficient of thermal expansion

Excellent thermal shock resistance

Great wear resistance

Very high fracture toughness

APPLICATIONS

Bearings

Mechanical parts

INFO

www.ceradyne.com

www.ortechceramics.com

ABOVE
A Silicon Nitride ball-bearing. If dropped on wooden floor, the bearing would leave a dent.

RIGHT
The NASA space shuttle bearings were made of silicon nitride. Image © NASA.

ALUMINIUM NITRIDE

An Aluminium Nitride wafer cutting through a piece of ice.

PROPERTIES

Thermally conductive

Stable

Electrical insulator

Piezoelectric

APPLICATIONS

Electronics

Dielectric optical storage

Military

Acoustic wave sensors

INFO

www.precision-ceramics.co.uk

Aluminium Nitride is an engineered ceramic characterised by its extremely high thermal conductivity, which gives the material its extraordinary heat sinking ability. Because of this, an Aluminium Nitride wafer held in the hand will cut ice like butter simply by the efficient conduction of the body's heat through the material.

It is stable at very high temperatures as surface oxidisation protects the material up to 1370 degrees Celsius or 2498 degrees Fahrenheit. Additionally it is extremely resistant to corrosion, making it ideal for precision ceramics and specialist military and aerospace applications.

PIEZOELECTRIC CERAMICS

PROPERTIES

Smart material

Moves when electrified

APPLICATIONS

Piezo-reactive fibres

Medical instruments for infertility treatment

Telecommunications

Piezoelectric motors

Actuators

Sonar

INFO

www.americanpiezo.com

Piezoelectric materials react dynamically to the application of an electric current, and conversely, produce an electrical charge upon the application of a mechanical stress, like hitting or bending. The Piezoelectric effect was discovered in the 1880s the pioneer crystallographer Pierre Curie. He found that certain types of crystals produced an electric charge when a mechanical pressure is applied.

Piezoelectric materials are commonplace in stove-top gas lighters and cigarette lighters, where small crystals are struck to create sparks. Some microphones are based on these materials too, where the changes in air pressure from incident sound waves vibrate the Piezoelectric material, which is then converted into an electrical signal.

Ceramics whose structures allow for the generation of piezoelectricity include lead titinate, litithium titinate and zinc oxide. Each one of these has different electromechanical properties, and can be used in specific applications where they are most appropriate, ranging from telecommunications to medicine. Given that these materials are able to convert pressure into electricity, their use for alternative energy production seems under-explored. Imagine a road capable of turning pressure from vehicle tyres into power.

MACOR

Macor is a white, Porcelain-like engineering ceramic comprised mainly of fluorphlogopite mica and Borosilicate Glass. It is used extensively in high-temperature environments such as lasers, nuclear reactors and aerospace applications because of its extremely good insulating properties and radiation resistance.

Macor can be engineered using standard working tools, allowing for it to be fixed using standard methods and fittings. In this way it is as versatile as an engineering plastic, but with the added benefit of being extremely heat resistant, able to be used in temperatures between 800 to 1000 degrees Celsius.

PROPERTIES

High temperature use between 800–1000 degrees Celsius or 1472–1832 degrees Fahrenheit

Highly insulating

APPLICATIONS

Spacers in lasers

Aerospace

Electronics

Superconductors

INFO

www.precision-ceramics.co.uk

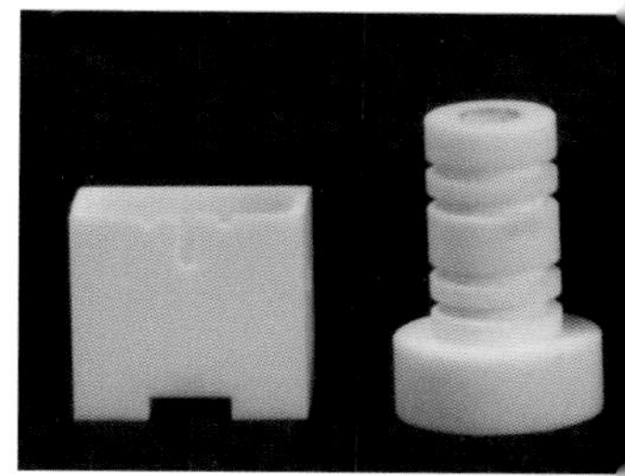

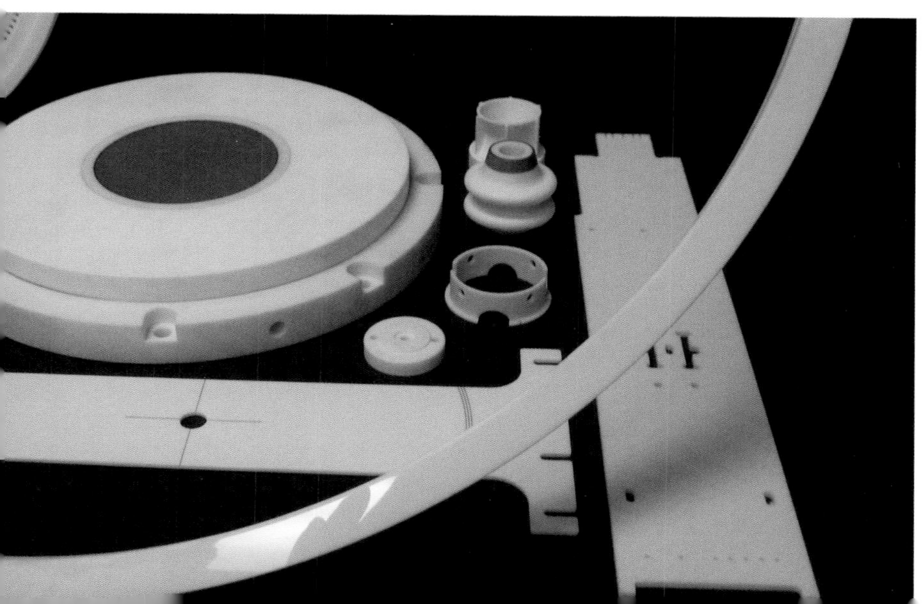

Technical parts made out of Macor. The material can be machined, enabling it to be engineered to make fine precision parts. Images courtesy Precision Ceramics.

CERAMIC BLADES

PROPERTIES
Strong
8.5 Mohs hardness
Non-magnetic
Non-conductive
Extremely sharp

APPLICATIONS
Knives

INFO
www.kyocera.co.uk
www.ziganof.com
www.edgeofbelgravia.co.uk
forever-k.comH

Ceramic Blades are made from a super-strong tough ceramic, often zirconium or zirconia. They are made by dry pressing zirconia powder, with subsequent firing through solid-state sintering. The resultant blade is sharpened by grinding the edges with a diamond-dust coated grinding wheel.

Zirconia ranks at 8.5 on the Mohs scale of mineral hardness, as compared to 10 for diamond. For this reason, the material lends itself to the manufacture of outstanding kitchenware—the hard edge of a Ceramic Blade seldom needs sharpening but if it chips it is not possible to re-sharpen it by grinding, unlike steel knives.

Ceramic blades are non-magnetic, non-conductive, and resistant to corrosion, whether in harsh environments or when exposed to caustic or strong acid substances.

LEFT
Zooming in to a Ceramic Blade shows how fine the texture of the material is, which, when sharpened, enables an even finer blade edge.

BELOW
Ceramic Forever, a brand of Ceramic knives for culinary use.

CERAMIC FOAM

Unlike some more familiar varieties of foam like sponges, Ceramic Foams are tough and rigid. There are two differing methods of manufacture that produce structurally different foams. The sol-gel method involves impregnating open-cell polymer foams internally with ceramic slurry, then firing the resultant combination in a kiln. The kiln burns away the polymer foam leaving only the ceramic. The other method is to 'foam' ceramic slurries by blowing gas through them, thus creating internal air bubbles.

The foamed ceramic slurries are commonly used as firebricks in kilns, as they are capable of withstanding very high temperatures. The most common usage for sol-gel manufactured Ceramic Foams is as filters, because they have very high internal surface areas and provide tortuous filtration routes for liquids to pass through. Both types of foams are also used for extremely high temperature insulation.

PROPERTIES

70–90 per cent porosity

1.0Mpa compression strength

Working temperature around 1500 degrees celsius or 2732 degrees Fahrenheit

APPLICATIONS

Industrial furnace lining

Acoustic insulation

Filtration

Substrate for catalysts

INFO

www.ceramic-honeycombs.com

www.ergaerospace.com

NOISE ABSORBING CERAMIC

Porocom is a noise absorbing material, made mainly from recycled ceramics. The name stands for porous construction material, and it is made from granules of recycled materials, such as clay and glass shard and sintered coal ashes, combined with a binding agent. The binder is made from a thermosetting powder coating (also an industry leftover). The material is heated to around 200 degrees Celsius or 392 degrees Fahrenheit, and the resulting chemical reaction is what creates the porous structure, and therefore the materials' insulating properties.

Not only is it made from 100 per cent waste materials, it can itself also be recycled. It reduces noise pollution and can be used in a number of applications, from fencing to ceilings and floors.

PROPERTIES

Recycled

Fire retardant

Good resistance to weather and UV

APPLICATIONS

Architecture

INFO

www.powdercoatingcentre.com

CASE STUDY

3-D PRINTING CERAMICS

THE WEDGWOODN'T TUREEN

The award-winning *Wedgwoodn't Tureen* harks back to the entrepreneurial role of celebrated potter Josiah Wedgwood during the Industrial Revolution. Now replacing the wheel with a computer, Cumbrian artist Michael Eden celebrates both past and new technological developments in creating ceramics. Known for innovative pieces inspired by both historical objects and contemporary themes, Eden's designs are highly stylised and transformative.

Eden's work was brought to life with Addictive Layer Manufacturing (ALM), also known as Rapid Prototyping or 3-D printing. In a move away from conventionally constructed (wheel-thrown or hand-built) ceramics and the limitations of "design for manufacture", Eden is freed from the material properties of clay classically worked by hand, making possible more complex shapes and intricate forms. Whilst professionals in electronics and manufacturing are using technical ceramics, reaping the benefits of what is now possible, designer-makers are likewise borrowing from industrial processes to create works that transcend what would traditionally be seen as 'craft'.

The *Wedgwoodn't Tureen* has been designed on Rhino 3-D and FreeForm software. It is produced by a ZCorp 3-D printing machine with the plaster material encased in a non-fired Ceramic coating. The recently formulated colours have ranged from loud yellow to striking pink. A famous relic reinvented for a digital age, the piece won an RSA Design Directions competition and one is now in the UK Crafts Council's collection.

OPPOSITE
GreyBloom and a detail revealing the complexity of the form, made from high quality Nylon material with a soft mineral ceramic glaze coating. No.1 in an edition of 24 in various heights and diameters. Image courtesy Adrian Sasson, London.

THIS PAGE
Twisted Oval Wedgwoodn't Tureen, 2011. Image courtesy Adrian Sasson, London.

VISCOUS PLASTIC PROCESSING

PROPERTIES

Increased fracture toughness

More complex and thinner shapes possible

Thin flat sheets can be made

Very smooth

APPLICATIONS

Technical ceramics

Plastic forming

Mass produced domestic crockery

INFO

www.molyandtantalum.com

Viscous Plastic Processing (VPP) is a process used to produce complex ceramic shapes that can easily be machined. Ceramic powders are mixed into a viscous polymer solution in tension, forming a kind of dough. This dough is then used to create complex shapes more commonly associated with plastics than ceramics. In fact, Viscous Plastic Processing allows ceramics to be worked in a similar way to plastics, such as by injection moulding.

Ceramics produced using VPP have considerably improved properties compared with other ceramics, virtually eliminating microstructural defects. Fracture toughness is improved, and the resulting product has far greater flexural strength than can be achieved by powder pressing.

VPP Ceramics are suited to applications where material reliability is critical, including aerospace, military and industrial components, as well as long lasting furniture and building tiles.

NEODYMIUM MAGNETS

PROPERTIES

Strongly magnetic

Permanently magnetic

APPLICATIONS

MRI Scanners

Pacemakers

Computer hard-drives

Headphones

INFO

e-magnetsuk.com

Neodymium Magnets are strong, permanent magnets made from the rare earth elemental metal neodymium, and are the strongest non-electric magnets currently available. These magnets are extremely difficult to demagnetise, which, coupled with their extreme strength, and makes them invaluable for use in science and medicine. They are also found in consumer electronics such as hard drives, electric cars and headphones.

Although they were discovered in 1885, Nodymium Magnets are being used increasingly frequently nowadays, acting as vital components in the electronic gadgets that we increasingly rely on more and more heavily.

HEXAGONAL BORON NITRIDE

Hexagonal Boron Nitride is a high-tech ceramic made by bonding boron to nitrogen. The "hexagonal" description refers to its structure, which is similar to Graphene, bonded in tight hexagons. Often referred to as "white graphite" because of its similarities to the material, Hexagonal Boron Nitride can be used in similar applications to graphite—particularly as a lubricant at extreme temperatures. It is most often used as a release agent for casting molten metal but because it does not react to the heat, it is also used as an additive to plastics to reduce their rates of thermal expansion.

Hexagonal Boron Nitride is an extremely fine, soft white powder, which only rates between one and two on the Mohs hardness scale. It was employed from the 1930s in the cosmetics industry as its texture lends it to use in compacts and foundations, however this died out in the 1980s as it became too expensive to produce. Nonetheless, recent changes in the methods of production have made the material much less expensive, and it is again a main constituent of cosmetics at all price ranges, from lipsticks to eye shadows.

PROPERTIES

Very fine 10 μm

Doesn't require water or gas particles

Used to 3000 degrees Celsius

Dielectric

Good thermal conductor

Mohs hardness 1-2

APPLICATIONS

Lubrication

Inclusion into Polymers

Cosmetics

INFO

ukabrasives.com

Hexagonal Boron Nitride is safe for use on human skin. It is therefore used as a base material for make-up such as eye shadows and powdered foundations.

SAFE AND CONDUCTIVE PAINT

BARE CONDUCTIVE

Bare Conductive have developed a range of non-toxic, electrically conductive materials, which they have now released in the form of two products; Bare Skin and Bare Paint. Both function as a conduit for electrical signals to run across surfaces, even skin!

The studio is a collaboration between four postgraduate students from the Innovation Design Engineering course at the Royal College of Art and Imperial College London. Bare Skin and Bare Ink have been trialled in their studio for such diverse applications as interactive henna-style tattoos, integrating LED components to greetings cards with simple circuits and kids' homemade toys with lights. One great application is as a safe and fun tool for children to learn about electronics.

Bare Skin is the only electrically conductive ink safe for use on the skin and is certified as a cosmetic in the EU, making it a great tool for theatrical make-up, extending the current range of possibilities for integrating electronics into costume and stage make-up.

Bare Paint is a non-toxic, multi-purpose material with a clearly demonstrated wide range of applications. Bare's website demonstrates how keen the studio are to work with designers, makers and hobbyists alike. The examples show the clear potential of this material with Arduino. The potentials for the safe integration of physical computing with the body in a playful way are huge. Beyond the arts and education, there is also potential for the material's use in medicine, technology and prosthetics. It doesn't contain toxic materials or expensive electronic components and can be used to print circuitry onto just about anything—a material with the ability to humanise electronics.

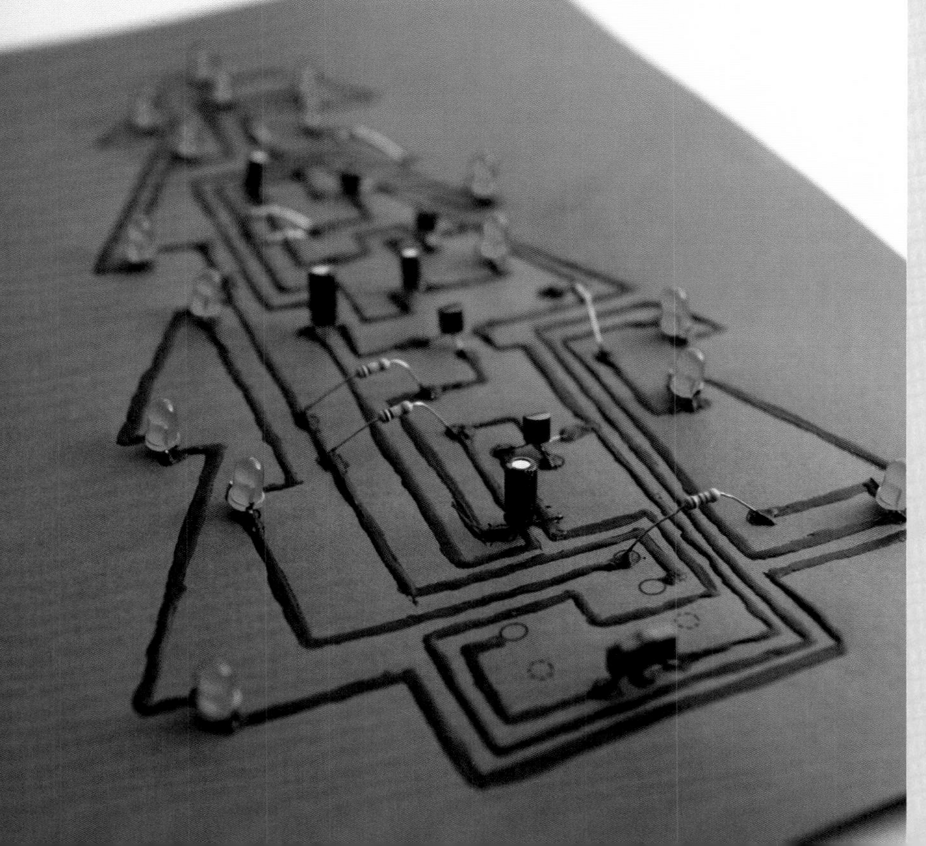

OPPOSITE
Bare Ink looks just like regular ink and can be used similarly. A printer can print a circuit on regular paper. The current will not affect the paper.

TOP
Bare Skin applied onto the skin to form interactive henna with LED lights.

BOTTOM
A novel Christmas card design which uses LEDs connected by Bare Ink. The ink is completely safe and non-toxic, despite being able to transmit a current. Images courtesy Bare Conductive.

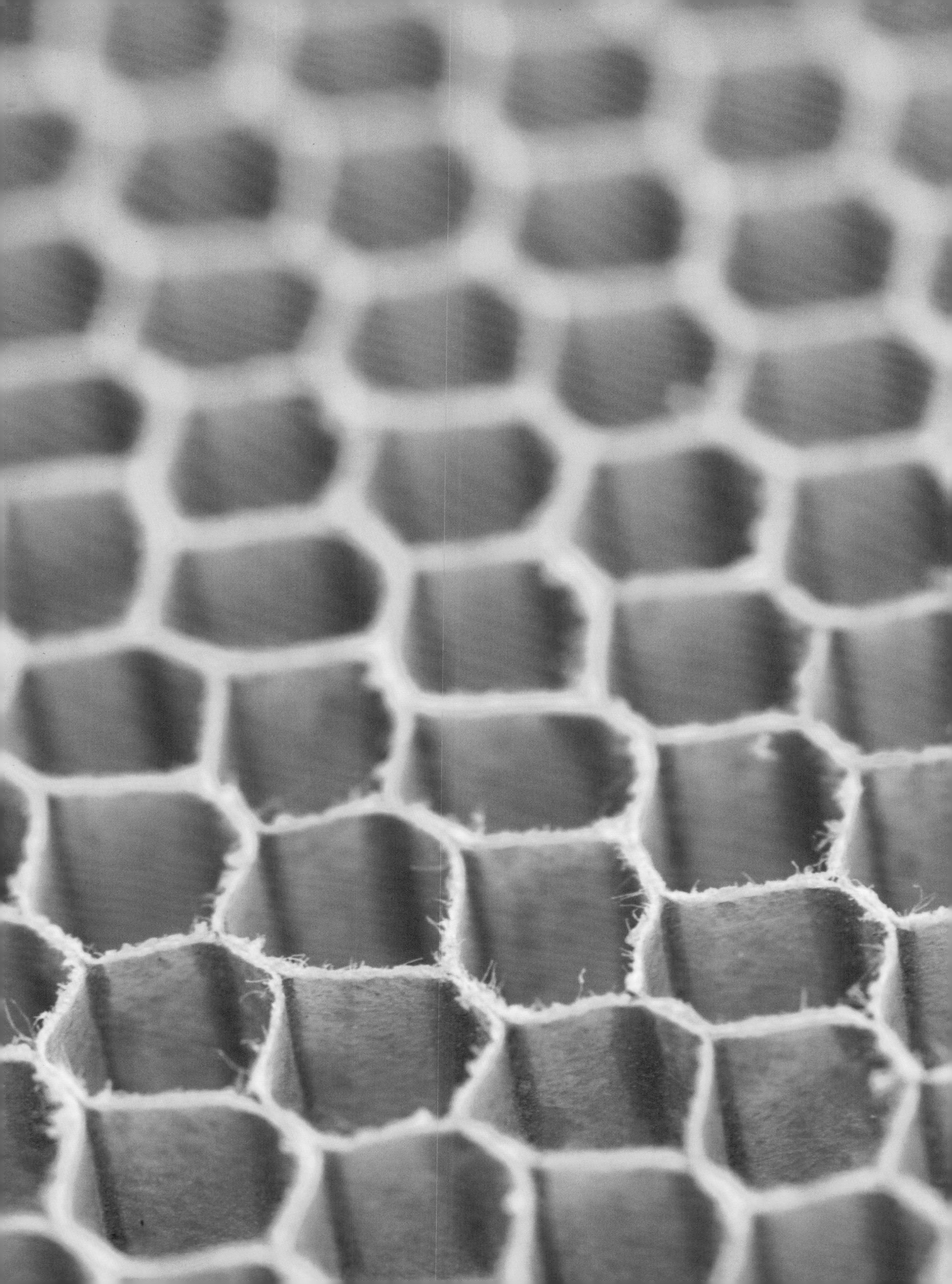

POLYMERS

Carbon is the basis of the molecular building blocks of life on Earth, a role which it plays because of its perfect chemical credentials. Carbon can make multiple bonds with neighbouring atoms, and forms long and complex molecules with all sorts of intricate branching and bending. In combination with the likes of hydrogen, nitrogen, oxygen, sulphur and phosphorous, it gives rise to a staggering array of chemicals, from the simple to the sublime. Importantly, carbon bonds can be formed and broken with a modest amount of energy, such that the complex organic chemistry that goes on in our cells can proceed with relative ease. However, this chemistry is not confined to the biological arena, as it is harnessed to make a class of material which is now ubiquitous—polymers. And in the same way that the ingredients of life are used to create a myriad of molecules and structures, the chemistry of polymers can be finely tuned to yield materials with a vast range of different and uniquely tailored characteristics.

Polymers are long molecular chains, consisting of a carbon backbone decorated with varying arrangements of hydrogen, oxygen, nitrogen, chlorine, fluorine and sulphur, and they are the building blocks of all plastics and rubbers. In contrast to the neat crystalline perfection exhibited within metals, polymers do no exhibit regular spacing and arrangement in their microstructure—they have an altogether more unruly architecture. Long polymer chains tangle into an intertwined mass, held together relatively weak electrostatic forces, or occasional strong chemical bonds. They are generally amorphous, although a degree of crystallinity can be obtained when chains line up alongside each other. The overall characteristics of plastics and rubbers are defined by the size and shape of these chains, and how they interact with one another when in close company.

Thermoplastics consist of linear polymer chains which are only weakly attracted to one another, but this is sufficient at room temperature to hold the chains together and make a solid structure. Elastomers are polymers (sometimes silicone-based instead of carbon) which give rise to elasticity in a material. Here, there is no attraction between the polymer chains, but there are occasional and localised points of strong chemical cross-linking between them. When the material is stretched, the individual chains can move around, slithering over and around each other as they stretch, however those occasional cross-links hold the structure together such that when the stretch is released, the chains snap back to their original positions. In contrast, thermoset plastics exhibit attraction between polymer chains and a large amount of strong cross-linking, such that the whole structure is very hard and inflexible, with the polymer chains fixed in place and unable to move. The vast numbers of polymers that can be made are matched by the wide variety of manufacturing techniques that can be employed to produce items from polymers; they can be extruded, blown and moulded into films, fibers, foams and much more, and can easily be coloured with no effect to the properties of the Polymer.

Although all plastics are polymers, not all polymers are plastics. Indeed, scientific definitions of polymers include woods and other organic materials that are made from long carbon chain molecules such as cellulose or starch. Such classifications are somewhat counterintuitive but help to demonstrate the breadth of polymers as a family of materials that encompass both the synthesised and non-synthesised material worlds. Over the years, the word plastic has even become a derogatory adjective used to suggest something is cheap or fake, be it the effect of too much make-up or the bad build quality of a child's toy. The rise of biodegradable plastics that derive their polymer chains from corn or rice rather than from petrochemicals perfectly demonstrates the new directions polymers are taking at a time when designers, manufacturers and consumers are becoming more aware of material life cycles and carbon footprints.

PH and ZL

PREVIOUS PAGES
Honeycomb paper made from aramid fibres.

LEFT
The ubiquitous coloured thermoplastics we so easily associate with the word "plastic".

CARBON FIBRE

PROPERTIES

Strong

Light

Good tensile modulus

Extremely fine

Easily incorporated into composites

APPLICATIONS

Aerospace

Sports equipment

Architecture

INFO

www.aircraftspruce.com

www.cstsales.com

Rarely seen as fibres alone because of their incorporation into composites, Carbon Fibres are in fact made from many chains of carbon atoms, and only five to ten micrometres thick. Single yarns are made up of thousands of these individual fibres wound together to form yarns. These yarns are then further woven to create the kinds of patterns that we are used to seeing in the bodies of racing cars or the shafts of golf clubs.

Made from fibres derived from petroleum, the Carbon Fibres are produced by isolating the carbon from the nitrogen in the fibre's molecular structure. Commercially, this is achieved through a process of heating the petroleum-derived fibre in an oxygen-free environment, which, because of the lack of oxygen, does not oxidise, leaving the carbon particles to vibrate and release the nitrogen atoms. The release of the nitrogen atoms causes the carbon to bond strongly.

Carbon Fibres were originally developed in the 1950s as a way to reinforce components on missiles. Nowadays, they are used for everything from components for spaceships to reinforcements for yacht sails.

ABOVE
A tangle of tiny individual Carbon Fibres, rarely seen in their bare form.

RIGHT
A Carbon Fibre under the microscope.

OPPOSITE
Designer Marc Newson worked with Carbon Fibres to create this ladder in a limited edition of 25. The ladder weighs just 2.2 pounds and can hold up to 220 pounds, such is the strength of carbon fibre. Marc Newson, *Carbon Ladder*, 2009, Galerie Creo. Dimensions: 48 x 38 x 201.5 cm. Photo credit: Fabrice Gousset. Image courtesy Marc Newson Ltd.

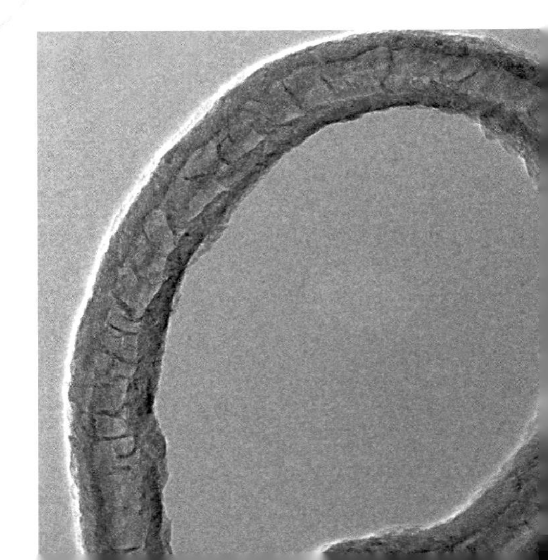

INJECTION MOULDED PLASTICS

PROPERTIES

Multiple parts produced from the same mould

Great variation

Quick process

Can use recycled materials

APPLICATIONS

Small parts

Bottles

Furniture

Toys

INFO

coolpolymers.com

www.ahrend.nl

www.yuyavsdesign.com

www.creativetools.se

Injection moulding is a highly efficient and swift way to produce a diverse range of plastic products, from disposable consumer items to specialist and intricate industry components. The simplicity of the injection process means that it is now essential in the production of plastic products in nearly all areas of manufacturing, particularly given its propensity for replacing more traditional materials, with a greater wealth of opportunity for creative freedom in designs.

The process works by way of a heated barrel equipped with a motorised screw feeding a molten—or plasticised—polymer in to a hollow split mould of the required product at high pressure. Derivatives of the basic process include multi-shot moulding, which incorporates different materials in the same mould; foam moulding, where the product has its density reduced by its material being 'foamed' in the mould; and insert moulding, where metal components are inserted.

As a particularly practical and innovative example of multi-part injection moulded manufacturing, the *XXXX STOOL* and *XXXX SOFA*—designed and made in collaboration between the Japanese designer Yuya Ushida and the Dutch firm Ahrend—comprises hundreds of replications of eight moulded elements. Different lengths of sticks, rings and joints were manufactured from recycled PET bottles and 'snapped' together to create an expandable and compressible whole. Yuya, previously a mechanical engineer and son of an iron-works owner, was inspired by the structure of the Eiffel Tower and bridges, a fact clearly visible in his idiosyncratic designs.

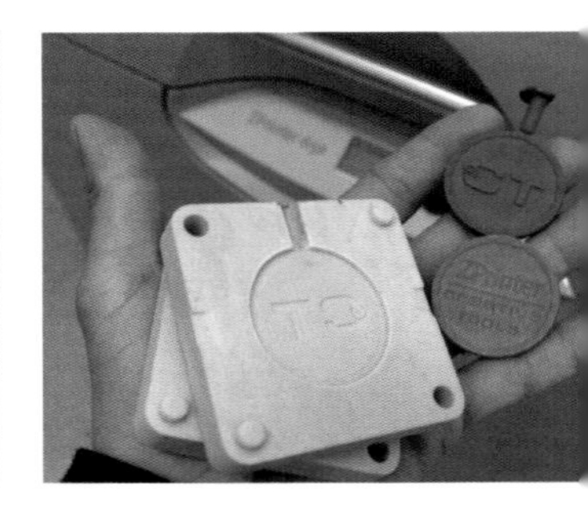

Combining computer aided design and scripting with rapid prototyping and injection moulding, Creative Tools, a Swedish technology company, have developed multiple and fast ways of reproducing shapes in plastics. Images Creative Tools.

Detail showing the components of the *XXXX SOFA*.

Yuya Ushida, *XXXX SOFA*. Using injection moulded parts connected together the chair shape expands to form a sofa. Images courtesy Yuya Ushida.

NYLON

PROPERTIES

Elasticity

Rigidity

High tensile strength

High-gloss

Available clear or opaque

Resin is injection—and compression—mouldable

APPLICATIONS

Fabrics

Meat wrappings and sausage sheaths

Carpets

Musical strings

Rope

Mechanical parts

INFO

www.hovding.com

Nylon was the first commercially successful synthetic Polymer and was exploited for its ability to be spun into silky fibres. This means it rapidly became a synthetic substitute for silk and has generally replaced it in applications such as hosiery and parachutes. Today, Nylon is a ubiquitous Polymer used in packaging, fabrics, carpets, and mechanical components. It even finds its way into the high-grade papers used in some currencies, making durable notes that can carry a variety of embedded anti-forgery devises.

Nylon is not simply one form of Polymer but a family of different types of polyamide. Such types, or units, of Nylon have names such as Nylon 6, Nylon 6/6, Nylon 6/11 and Nylon 11, to name but a few that are commonly used in industry today. Each of these different types possess different properties and as the numbers of the units increase, the structure and resultant properties of the material become more similar to polyethylene—that is they become less strong, less hard and have lower melting points. Engineering-grade Nylon is processed by extrusion, casting, and injection moulding.

The properties of Nylon can also be enhanced or altered when combined with other materials to form Nylon-based composites. Adding glass fibres, for example, can increase the strength of the Nylon, and specific minerals can enhance chemical-resistance.

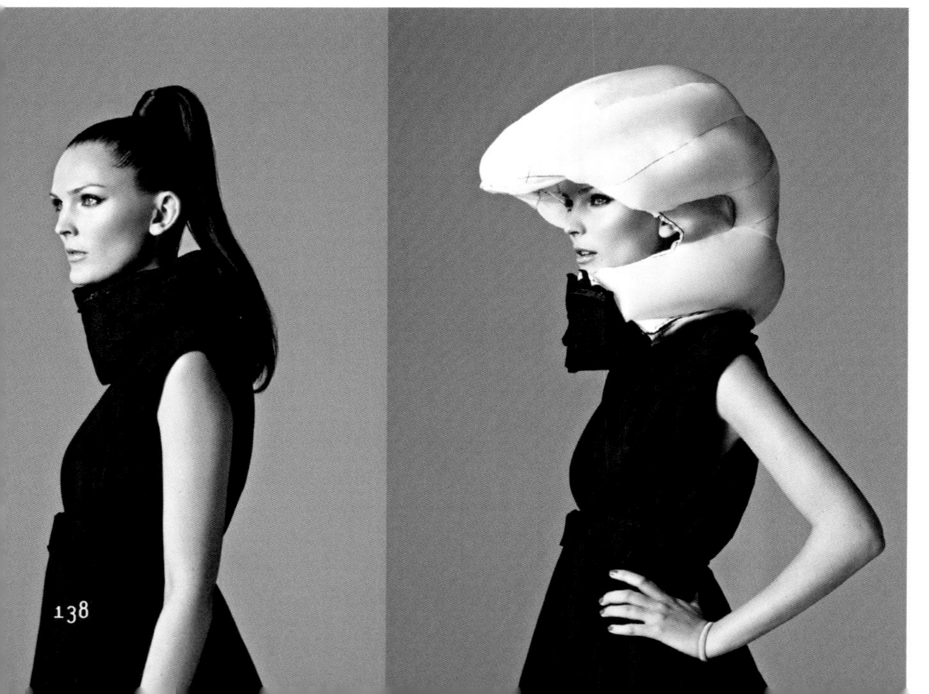

Because Nylon is extremely strong yet flexible, Hövding use it to create their inflatable bike helmet. The design folds up small enough to fit in a small wearable collar which, when triggered by unusual cycling motions determined by extensive testing and sensors, deploys the helmet in a fashion akin to automobile airbags. The helmet is worn as a collar and only deploys when required. Using nylon enables the bag to be folded into a small pouch in the collar yet be strong enough to withstand a serious accident. Images courtesy Hövding.

ADDITIVE LAYER MANUFACTURED POLYMERS

The Additive Layer manufacturing process works by fusing together layers of plastics, Nylon or powdered metal, melting the material bit by bit, and then allowing it to solidify to the requisite shape, before adding more and more layers to strengthen and complete the form. Engineers at the European Aeronautic Defence and Space Company (EADS) built a prototype space-age bicycle that was constructed from Nylon but was as strong as steel. In this way, the component sections of the bike have been built as only six single pieces, including some moving parts in the wheels—such as sealed bearings within the bicycle's wheel hubs. The frame's stiffness is maintained through its integrated truss structure, and there is even a level of saddle comfort, achieved through a printed auxetic structure.

The Airbike is so-called due to the fact that Airbus was the first EADS company to embrace the process. Its purpose is largely demonstrative; the real potential of the construction method, according to EADS, is in "high-stress, safety critical" aviation use. The technology is hypothetically very green as well, for the resultant reduction in vehicle weight would have a direct effect on the amounts of fuel required to power the vehicle.

PROPERTIES

Strong

Cheap to manufacture

Hard

APPLICATIONS

Building complex forms

Mechanical parts

3-D scanned object copies

INFO

www.eads.com

Airbike MkII as presented by Jean Botti (EADS Chief Technical Officer) at EADS' media seminar in the run up to the Paris Air Show 2011.

POLYURETHANE

PROPERTIES

Abrasion resistant
Oil and solvent resistant
Weather resistant
Load bearing

APPLICATIONS

Casting
Toys
Furniture
Model-making

INFO

www.sdplastics.com
www.no-smoking.com

Polyurethane comes in a vast variety of forms varying in rigidity and density, from hard casting plastics used for tightly sealing liquids to squishy foams you can sit on to epoxies and every consistency in between. It is often used for applications where strength and flexibility are required, such as hoses, because it combines the best qualities of rubber and plastic.

A Polyurethane is a polymer comprising a chain of organic units joined by carbamate (urethane) links, from which it gets its name. The variations in the consistency and properties of the materials depend on the organisation of the chains of units. The combining process usually requires a catalyst.

The German designer Christof Schmidt developed a new way of joining wood together using Polyurethane. In his chair titled *DaR*, the wooden components are first broken and set in a mould that is then filled with polyurethane. Because the polyurethane expands when it sets, it finds its way into all of the broken wood fibres, securely holding the pieces together.

ABOVE
The wood is broken and the pieces are set in a mould into which liquid Polyurethane is poured.

OPPOSITE
DaR chair by Christof Schmidt.

RIGHT
Detail of the Polyurethane joint showing the way in which the material has flowed into the cracks and solidified. Images courtesy Christof Schmidt. Photographs © Ali Schmidt.

CASE STUDY

CASTING FROM FLAT SHEETS

FACETURE VASES

Designer Phil Cuttance's project *Faceture* uses numerous Polymers to create a series of individual vessels, light-shades and tables. By casting a water-based resin into a simple hand-made mould, Cuttance is able to manipulate each mould before casting, making each piece unique. In making the most of the properties of the materials, he has created a variety of shapes and sizes from the one process.

Using a simple sheet of 0.5 millimetre plastic scored, folded and taped into shape, Cuttance was able to produce a mould that can be varied for each cast. He tapes it each time for a different result, with some of the triangles popping inwards and some popping outwards. He writes "I do this each time I ready the mould for the next object, meaning that no two castings are the same."

Using a water-based casting resin allows a quick working time. The resin can simply be poured into the hollow mould and rolled around to coat the inside, controlled by Cuttance on the casting jig. He repeats the process using another colour, which means he can have different interior and exterior colours, further increasing the variations. The resultant products have a sharp digitally rendered look, which is a surprise given their 'low-fi' methods of creation. Each vase is hand-made, unique, and numbered on the base.

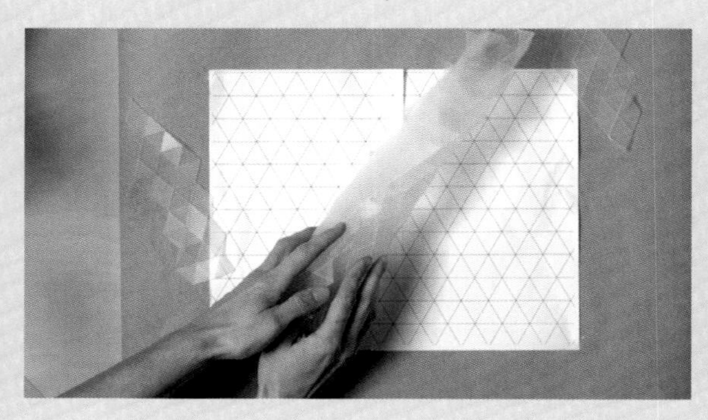

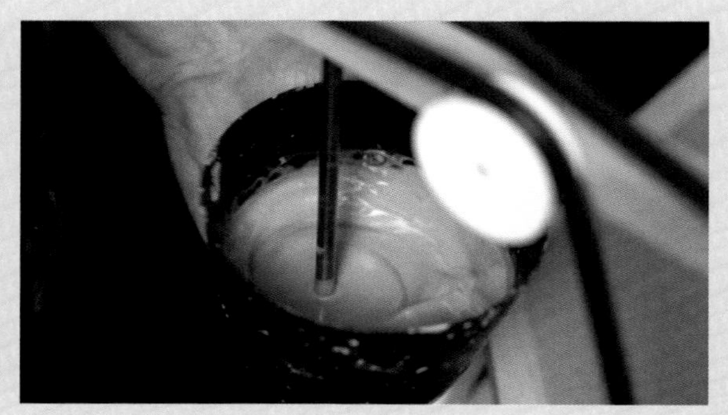

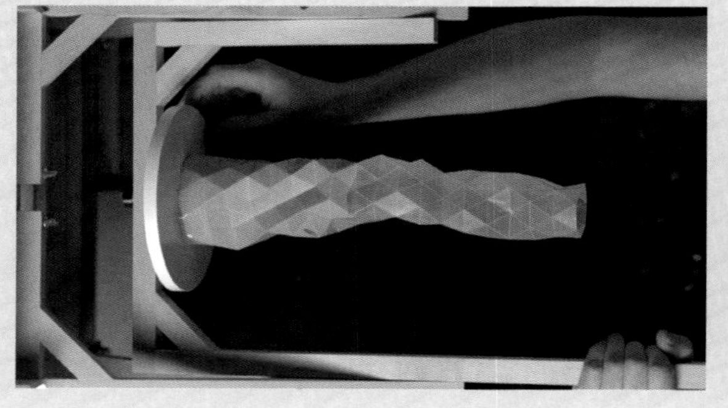

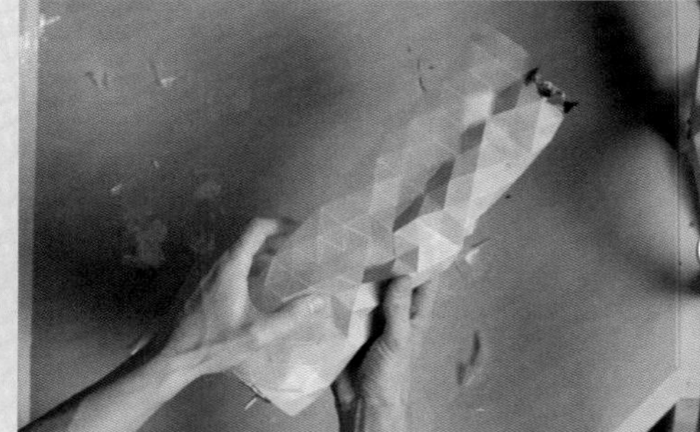

OPPOSITE TOP LEFT TO BOTTOM RIGHT
The sequence of making the vessels. Folding the scored mould. Stirring the liquid plastic. Filling the mould on the Faceture machine as seen from above and taking off the mould.

THIS PAGE
Faceture vases, each cast in two colours—a darker colour outside and a a lighter colour inside. This is done by coating the mould twice. Images courtesy Phil Cuttance.

POLYSTYRENE

PROPERTIES

Easy to form

Light-weight

Long life

APPLICATIONS

Beco building material

Packaging

Filler material

Soil aerators

INFO

www.becowallform.co.uk

Polystyrene is a sweet-smelling polymer made from the organic compound Styrene, itself a liquid hydrocarbon created in the chemical industry from petroleum. It can be utilised in product manufacture as a solid plastic—as in its application for disposable cutlery, smoke detectors and CD casings, for instance—or the more synonymous 'foamed' use, resulting in the ubiquitous box packaging blocks and takeaway cups.

Polystyrene can be either a thermoset or thermoplastic Polymer. The former is a Polymer that, once liquefied by heat, remains in its re-set cooled from permanently; the latter can be melted—specifically in this case at around 100 degrees Celsius or 212 degrees Fahrenheit—and remoulded again and again.

Beco's WALLFORM is a building system made from Expanded Polystyrene. Its method of use comprises assembling a myriad of hollow interlocking Polystyrene blocks, which, when in place, are filled with concrete. When this hardens, the surrounding blocks can then be cut down to the requisite shapes and thicknesses. External claddings, such as wood, masonry and render can be added, and upper floors and roofs are easily applied. It is even possible to build tall structures this way, and once rendered, one would never guess that the majority of the form is the same material as take-away coffee cups are often made from.

NOT-SO-EXPANDED-POLYSTYRENE

PROPERTIES

Solid

Structurally sound

"Noisy pixelated quality"

Material records the textual quality and stitching of the fabric mould

APPLICATIONS

Furniture

INFO

www.silostudio.net

The *Not-So-Expanded-Polystyrene* (NSEP) process, devised by Royal College of Art graduates Attua Aparicio and Oscar Wanless' Silo studio, is an innovative use of Polystyrene, creating a solid and structurally sound new material.

The NSEP process involves making furniture by steaming Polystyrene beads inside fabric moulds. This causes the beads to melt, expand and fuse together in a constricted space. This creates solid shapes and blocks of material that record the textural quality and stitching of the fabric mould. Different coloured grains are used to achieve—in the designers' words—a "noisy pixelated quality" within the "dark and playful" sausage-like aesthetic of the furniture pieces. The pair have recently set-up their SILO Studio in the corner of a plastics factory, developing their method as part of a residency, locating materials experimentation and design processes at the site of commercial manufacture.

LEFT
SILO Snot.

RIGHT
SILO Cave and *Boulder*.

ABOVE
Table detail.

LEFT
SILO Chair with *SILO Table* and *SILO Shelves.*

RIGHT
SILO Chair.

BELOW
SILO Studio Table. Images courtesy SILO Studio.

The *Not-So-Expanded-Polystyrene* works by SILO Studio's Oscar Wanless and Attua Aparicio.

POLYETHYLENE TEREPHTHALATE [PET]

PROPERTIES

Recyclable

Can be made extremely clear

Easily formed

APPLICATIONS

Food and drinks containers

Fibres

INFO

www.rosslovegrove.com

www.napcor.com

PET is a safe, recyclable plastic most commonly used for containing fluids. It is confusing because it doesn't actually contain Polyethylene, therefore when refered to in science it is often called Poly (Ethylene Terephthalate).

Depending on how it is processed it can exist as both an amorphous transparent and a semi-crystalline Polymer. Most often seen as cloth in synthetic fibres, the name PET is best known as the material of choice for water bottles. However, when used for fibres it is known as Polyester.

With a resin identification code of 1, PET bottle recycling is the most practical of thermoplastic recycling. Most of the plastic bottles we recycle become Polyester fibres for fleeces.

Although PET is regularly used for water bottles, Ross Lovegrove's redesign of the Ty Nant water bottle was the first to really make use of the material's aesthetic qualities and similarities to water. The fluid design captures the clarity and movement of the water it contains.

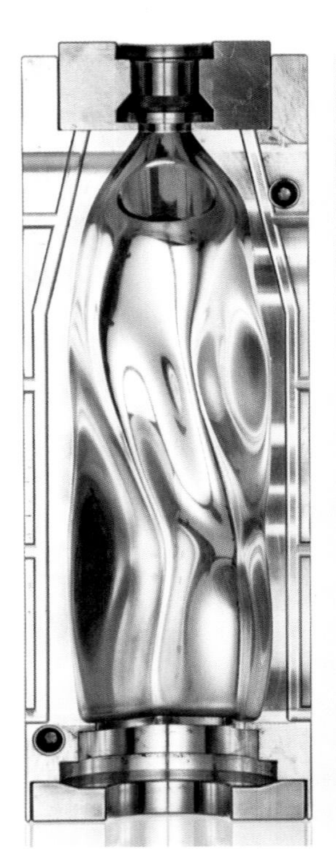

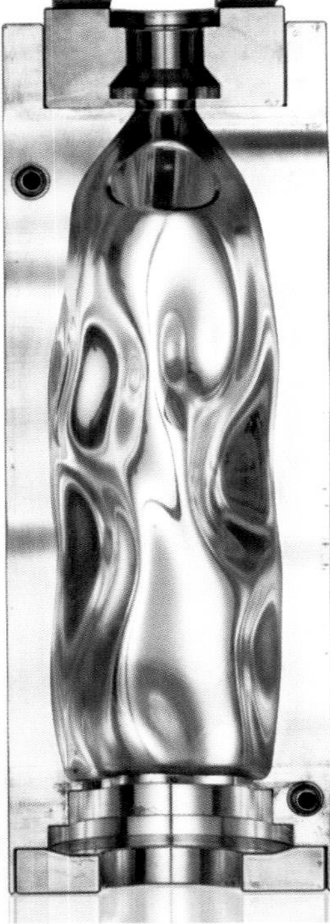

LEFT
PET is blown into the moulds, creating a thin layer of plastic on the surface. Image courtesy Ross Lovegrove.

OPPOSITE
Ross Lovegrove's design for the Ty Nant water bottle, half filled with water. The PET captures its contents' fluidity. Image courtesy Ross Lovegrove.

TOP LEFT
Patagonia's Common Threads Initiative in action: worn-out clothes ready for recycling. Patagonia Distribution Center, Reno, Nevada. Photograph, Jeff Johnson.

TOP RIGHT
Way too much of what is made these days ends up in landfill at the end of its useful life. At Patagonia, they're trying to change this pattern. Photograph, Pete Ryan.

CENTRE
Detail of clothing awaiting repair at Patagonia's repairs department.

RIGHT
Thread that has been converted from recycled Capilene at the Teijin facility. Images courtesy Patagonia.

RECYCLED POLYESTER

Patagonia, an outdoor clothing manufacturer, have been making post-consumer recycled plastic soda bottles into clothing since 1993. Their website states: "over the course of 13 years, we saved some 86 million soda bottles from the trash heap. That's enough oil to fill the 40-gallon gas tank of the diminutive Chevy Suburban 20,000 times". The implication is clearly that a lot of oil is saved in this act of recycling, though less information about the energy used or saved is present.

Recent work with the chemical research company Tenjin has enabled Patagonia to establish a direct clothes recycling system, where any Polyester garment, such as a thermal fleece, can be recycled to produce new clothes.

PROPERTIES

Lessens dependence on oil reserves

Creates less air, water and soil contamination

Curbs discards, thereby prolonging landfill life and reducing toxic emissions from incinerators

Helps to promote a new recycling stream for polyester

APPLICATIONS

Clothing

Recycling plastic soda bottles

INFO

www.patagonia.com

ww.tejin.co.jp

Used garments provide fabric patches for Patagonia's repairs department. Patagonia Distribution Center, Reno, Nevada.

CASE STUDY

MATERIAL FOR EVERYTHING

SUGRU

Whilst working for her masters in Product Design at the Royal College of Art, Jane ni Dhulchaointigh decided to dedicate her project to anti-consumerism, and, in the process, began to develop a material capable of mending and amending commercially bought products to suit the needs of the user. After her studies she continued to develop the material, experimenting with composition and developing a brand that celebrated the potential of the material as a way of hacking objects.

Sugru is adhesive to most materials, such as aluminium, ceramics, cotton, glass, acrylic, steel and wood. The name is inspired by the Old Irish word for 'play' and indeed the product looks like Play-Doh. Once out of its packaging it can be moulded easily by hand for up to half an hour. Twenty-four hours later, the rubber material has air-cured to form a strong silicone, which by nature is flexible and waterproof. It is practical, inexpensive and easy to remove.

LEFT
A camera is made child-friendly with Sugru. Once cured, the material remains bouncy, meaning that if dropped this camera will bounce rather than break.

OPPOSITE LEFT
Sugru comes in sealed packs that, once opened, have a 45 minute working time.

OPPOSITE RIGHT
Sugru can be used to mend or amend just about anything, from car keys to headphones. Images courtesy Sugru.

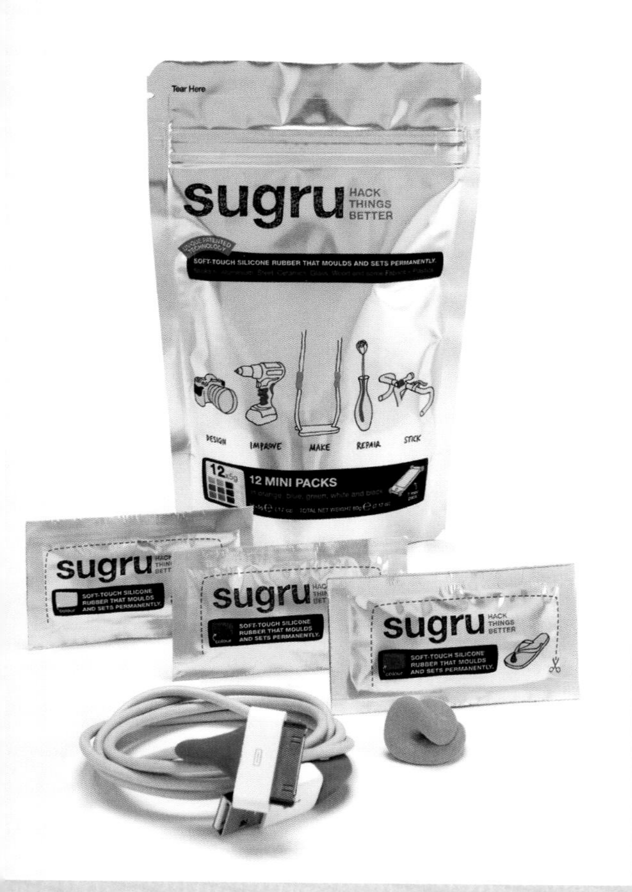

Colours come in a variety which can be mixed together to create almost any conceivable colour, so that the material can be tailored to the requirements of the user. Some choose to make the Sugru addition in bright colours, drawing attention to their hack action, whilst others match the colour of the Sugru to the object, making the addition less obvious. It is proving a hit with design and DIY enthusiasts, enabling them to personalise tools, toys, electronics, and the environment around them.

The product is notable for its durability in various extremes, be it temperature, pressure or wet conditions. It is environmentally friendly, harking back to the days when 'make do and mend' was in the public consciousness and materials were in repair rather than in landfills. From repairing the inside of a washing machine to adorning walking boots in the great outdoors, there is not much that Sugru cannot do; it is limited merely by the owner's imagination.

RECYCLED PLASTICS

PROPERTIES

Less wasteful than creation of new plastics

Huge variation achieved

Can simply melt plastics and keep their former identities

APPLICATIONS

New plastic products

Up-cycled products

Counter-tops

Panels

Clothing

INFO

www.smile-plastics.co.uk

www.durat.com

www.evd-diez.de

www.emtex.huVaria

www.3-form.com

There is plenty of opportunity to recycle plastics, the most obvious being those plastic products that can be used again in their original form—such as milk cartons. Thermoplastics in particular offer good recycling opportunities, due to their specific behaviors when exposed to heat, as they can be melted down to form stock that can be reused in other plastic manufacturing processes.

The quality of Recycled Plastics depends upon the stock plastics used and how specifically they have been sorted. Blends can be created from various thermoplastics to make thick-walled thermoplastics, but many applications require a purer Polymer stock, thus calling for advanced sorting of the recycled materials. Recycled Plastics are not preferable, however, in the production of high-performance plastics, due to the loss of quality in the material.

Duroplastic resin systems and elastomer plastics offer very limited opportunities to recycle due their needing to be chemically converted into basic components.

RIGHT
Detail of red sink. The texture of the recycled material that comprises Durat can be seen.

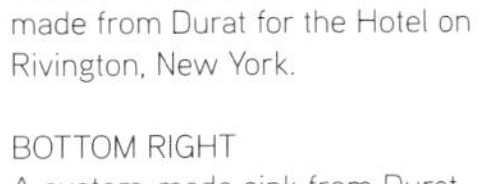

BOTTOM LEFT
Custom wash basin and custom bath made from Durat for the Hotel on Rivington, New York.

BOTTOM RIGHT
A custom-made sink from Durat in yellow. Images courtesy Durat.

SONIC FABRIC

Sonic Fabric was devised by conceptual artist Alyce Santoro from woven polyester thread and recycled audiocassette tape. Because the fabric is woven from cassette tape, which itself functioned because it was embedded with magnetic ferric oxide, when touched with an electromagnet the resulting fabric can be 'played'. The weaving creates a distorted sound of different notes.

When made into garments the clothing can be performed, but the inherent weave, texture and materiality also make it an interesting fabric in its own right.

PROPERTIES

Can be played using electromagnet

Recycled

APPLICATIONS

Performative clothing

INFO

www.alycesantoro.com

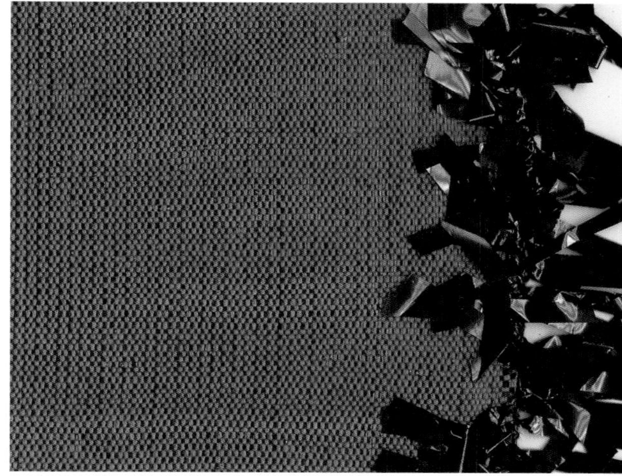

TOP LEFT
Sonic Fabric *Voidness Dress*, 2007. Photograph by Erik Gould, courtesy of Rhode Island School of Design Museum.

TOP RIGHT
Three Sonic Fabric Neckties. Photograph by Alyce Santoro.

BOTTOM
Detail of Sonic Fabric showing the weave pattern of recycled cassette tape. Image courtesy Alyce Santoro.

RUBBER

PROPERTIES
Very elastic
Very resilient
Can age and become brittle

APPLICATIONS
Airbus A380 tyres
Condoms
Balloons

Extracted from the sap of the rubber tree, Rubber consists of water and sulfur and is characterised by its enormous bounce when hardened.

It is a naturally occurring elastic hydrocarbon polymer. It is so resilient as a material that it is used in almost all Rubber blends for tyres. Other uses include flooring, hoses, balloons and condoms. Wet Rubber can also be used as an adhesive.

Rubber doesn't dissolve easily in solvents. To fully break it down it must first be shredded then immersed in a solvent such as turpentine.

CHARCOAL FILTER

PROPERTIES
Removes impurities
Adds nutrients

APPLICATIONS
Purification Water
Soap
Shampoo

INFO
www.sortofcoal.com

Charcoal is an extremely good filtration material. It is used extensively in industrial applications, and is the main filtration material in ecologically friendly water filters that you might find in bottles or jugs. White Charcoal is a type of traditional Japanese charcoal derived from oak that is celebrated and utilised for its excellent purifying qualities with water and air, removing impurities and adding nutrients.

The company that manufactures White Charcoal, Sort of Coal, offers a range of products made from the material, from soap to water purifying charcoal sticks, which remove impurities and add nutrients to water. They only need to be in contact with the water, rather than having it pass through, to have an effect, meaning sticks of it can simply be placed in jugs of water.

LEFT
White Charcoal filtering water through immersion in a container. Image courtesy Sort of Coal.

RIGHT
The slow burning causes a dense texture of cracks, perfect for filtration. Image courtesy Sort of Coal.

MUSHROOM PACKAGING

Developed by Ecovative Design, Mushroom Packaging is made from agricultural byproducts, such as cotton seed, wood fibre and buckwheat hulls. These are combined with mycelium to develop a fungal network that effectively bonds the agricultural byproducts together within the root network of mushrooms. The funghi subsequently digests the agricultural waste, binding it into a structural material, much like glue.

Intended to be used as an alternative to foam packaging, Mushroom Packaging is 100 per cent renewable and home-compostable.

PROPERTIES

Renewable

Home compostable

APPLICATIONS

Packaging

INFO

www.mushroompackaging.com

These samples of Mushroom Packaging can be 'grown' and used to replace traditional Polystyrene in packaging of electronics. Images courtesy Ecovative.

MAKING YOUR OWN PLASTIC

A WORKSHOP FOR DESIGNERS

Swiss furniture and product designer Beat Karrer ran a workshop in conjunction with Vitra experimenting with bio plastics to assess new applications for the material. Bioplastics can be produced from raw natural materials like cellulose, protein or starch, each of which can be extracted from plants or animals. Beat Karrer's team conducted a workshop teaching designers the process of making samples of Bioplastics and realising their potential to help combat the waste crisis.

They made Bioplastics from potatoes by peeling 30 kilograms of spuds in order to extract the starch needed to start the procedure. Another material they experimented with was Biograde, a polymer based on cellulose extracted from vegetables that is resistant to extreme temperatures. It has also been approved for use as food packaging and is biodegradable.

Although ideal for products such as packaging with a short lifespan, Bioplastics are less commercially popular than their cheaper counterpart. Increasing costs of oil and diminished resources of other fossil fuels are however alerting the public to new ways of developing plastics. Workshops such as these help designers to understand how these new materials are made, and can therefore help them to understand what their potentials are.

Some experiments with Bioplastics made from cellulose and potatoes during the workshop. Images courtesy Beat Karrer.

TOP LEFT
The designers in the workshop had to peel 30 kilograms of potatoes in order to make a starchy Bioplastic. Image courtesy Beat Karrer.

TOP RIGHT
A potato starch gluey mixture—the basis of a do-it-yourself Bioplastic. Image courtesy Beat Karrer.

BOTTOM LEFT AND RIGHT
Experimenting with Bioplastics has led Beat Karrer to develop FluidSolids®. His Stool and Bowls have received the Materialica design and technology Gold award for CO_2 efficiency in Munich. Image courtesy FluidSolids®.

BIOPLASTIC

PROPERTIES

Made from completely renewable resources

High content of naturally sourced material

Impermeable to humidity/gas transfer

Breathable

Tough

Biodegradable

Compostable

Easy to colour, print and extrude

Wide processing window

APPLICATIONS

Packing materials

Replacement for conventional plastics

INFO

www.fkur.com

www.natureplastics.com

Bioplastics are made using renewable biomass sources such as cornstarch or vegetable oil as an alternative to the traditional use of non-renewable fossil fuels.

There are many different types of Bioplastics. Some are made from completely renewable resources, others with just a percentage of them included. BIO-FLEX, for example, is made from Co-polyester and PLA, with a high content of naturally source material. Such plastics are most often used for packaging, and need to be formulated with various properties to suit different applications. They will often have to be impermeable to humidity or gas transfer, but will sometimes have to be breathable while still remaining tough. Creating Bioplastics with such varying functions requires much experimentation with processing and formulations.

Bioplastics are mainly used to replace conventional plastics such as low density Polyethylene, high density Polythene as well as Polystyrene and Polypropylene. As such, they are able to be used for a number of products we take for granted such as packing materials, which, often made of pure Polythene, are extremely wasteful as they are rarely used again. For example, BIO-FLEX is considered better for the environment than Polythene as it can biodegrade and is therefore compostable, but is just as easy to colour, print, extrude and has a wide processing window.

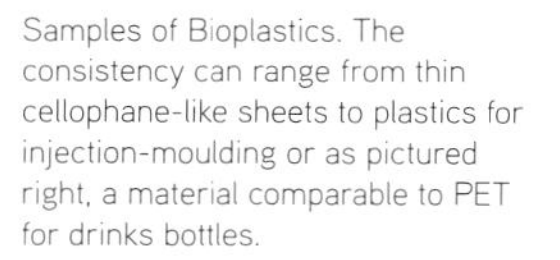

Samples of Bioplastics. The consistency can range from thin cellophane-like sheets to plastics for injection-moulding or as pictured right, a material comparable to PET for drinks bottles.

POLYLACTIC ACID (PLA)

Polylactic Acid, otherwise known as Polyactide (PLA), is a 'biocrude plastic' produced from renewable resources, either through the fermentation of sugar syrup or the bacterial fermentation of starch or other sugars.

Comparable to Polyethylene Terephthalate (PET), it is often colourless and shiny when in solid form. PLA's properties are dependent on its composition—for example, by adding fibres, its mechanical resistance can be improved. Used widely in the packaging industry, PLA is expected to eventually overtake the use of PET due to its biodegradability and the fact that it can be made from sustainable crops, unlike PET. A disadvantage is that it softens at very low temperatures.

PROPERTIES

Low permeability for gases

Water repellent surface

Low-heat stability

APPLICATIONS

Medical applications

Automobile industry

Entertainment industry

Geo-textiles

INFO

www.natureworksllc.com

www.bioplastics.basf.com

www.fkur.com

www.gehr.de

www.earthdistributors.com

Alstrom BioWeb™ teabags, made from PLA. Image courtesy Fkur.

WATER SOLUBLE PVA (PVOH)

Water-Soluble Polyvinyl Alcohol (PVOH) is a water-soluble plastic that has thermoplastic properties. Whitish-yellow in colour, PVOH has excellent adhesive properties, great tensile strength and flexibility. The plastic works by absorbing water, which makes it soft before the water eventually dissolves it. As with other thermoplastics, PVOH can be extruded, injection molded, and thermoformed to make various different products.

PVOH is mainly used in packaging, for water-soluble products, such as bath pearls and as an adhesive and thickening agent in products such as shampoos, glues and hairsprays. The paper industry also uses PVOH as a binding agent.

PROPERTIES

Adhesive

Great tensile strength

Flexible

APPLICATIONS

Toiletries

Paper

Glue

Textile production

Metal injection molding

NEWSPAPER WOOD

Designer Mieke Meijer, having pondered the traditionally irreversible process of making paper from trees, developed Newspaper Wood. Aesthetically and practically resembling real wood, it can be chopped or sanded. Newspaper Wood looks like wood, but is still a paper product. The newspaper is stacked, glued, then rolled into a log shape and compressed until dry. The glue is free of solvents that would make recycling difficult.

Meijer's project exhibits the trademark clear lines and fine detailing of Vij5's designs. Vij5 is a Dutch design label, founded by Anieke Branderhorst and Arjan van Raadshooven, characterized by simplicity and the use of existing elements and "pure, honest materials". In April 2011, the first prototypes were presented in Milan, exhibited at the Milano Design Week. So far operating on a small scale the project's main aim is to generate "upcycling" and be valuable in other contexts.

PROPERTIES

Recycled

Recyclable

Resembles wood

APPLICATIONS

Furniture design

Interior design

INFO

www.vij5.com

OPPOSITE
The rolled and glued sheets of newspaper, when sliced through, look similar to grains of wood.

THIS PAGE
Cutting the newspaper wood into in planks makes it usable in furniture design. The first prototype collection was designed by rENs, Breg Hanssen, Greetje van Tiem, Ontwerpduo, Floris Hovers, Christian Kocx & Tessa Kuyvenhoven. Images courtesy Vij5.

WOOD POLYMER COMPOSITES

PROPERTIES

High rigidity

Tensile strain at break
0.3–1.3 per cent

Good acoustic properties

Made from renewable material

Good replacement for crude oil

Yield Stress between 18-61 N/mm2

APPLICATIONS

Press moulding complex shapes

Car interiors

Handles

Cases for hand-held consumer electronics

INFO

www.tecnaro.de

Often called "Liquid Wood" because of its ability to be melted and used in thermoplastic processes, taking on new three-dimensional forms, Wood Polymer Composites (WPCs) consist of both wood fibres and a plastic matrix.

Wood Polymer Composites are useful for press moulding complex shapes such as car interiors and handles or cases for hand-held consumer electronics, but because of the high wood content (usually between 50 and 90 per cent) they cannot be processed at a temperature above 200 degrees Celsius or 392 degrees Fahrenheit.

There are many different variations of Wood Polymer Composites, with different variations and visual effects, from those that visually resemble different wood shades to plastics that hardly resemble wood at all.

INSULATING SEAWEED

PROPERTIES

Doesn't rot

Found naturally

Inexpensive

Naturally fire resistant

APPLICATIONS

Insulation

INFO

www.neptutherm.de

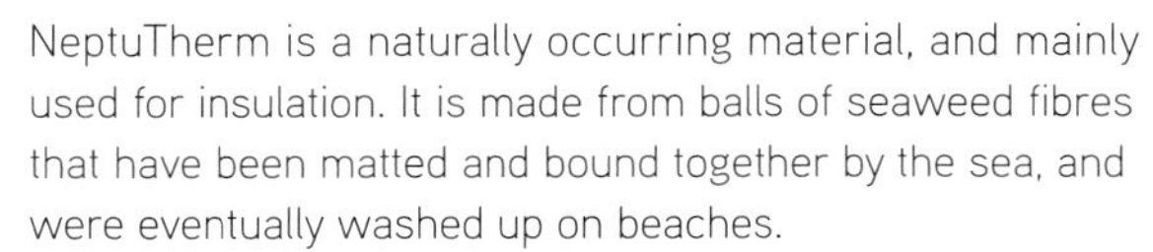

NeptuTherm is a naturally occurring material, and mainly used for insulation. It is made from balls of seaweed fibres that have been matted and bound together by the sea, and were eventually washed up on beaches.

Low in salt and protein, it does not rot and is therefore safe for residential use, particularly when combined with its natural fire prevention properties.

The seaweed washes up on the beach as fibrous balls, which when dried are used as very effective insulation either shredded or in their found form. Image courtesy NeptuTherm.

CORK POLYMER COMPOSITE

Comprised of cork particles, Cork Polymer composites are largely made up of suberin, a waxy substance that is the main constituent of cork. It is this complex Polymer that renders the material impermeable to water vapour. CPCs are bonded by a thermoplastic material, which, as the name suggests, are shaped by heat thereby combing the qualities of natural cork with the resin binder, lending strength to the cork's flexibility.

The bio-based material proves that cork doesn't just belong to notice boards, trees and wine stoppers; a wide range of cork products, from helmets to suitcases, are on the market. It can be manufactured by extrusion or injection moulding machinery.

PROPERTIES

Flexible

Noise and vibration-reducing

Based on renewable raw materials

APPLICATIONS

Architecture—wall panels in the building industry

Furniture—lamps/vases/wash basins

Medical e.g. orthopaedic soles

Sport—bicycle and ski pole handles/sport mats/shoe parts

INFO

www.amorim.com

www.simpleformsdesign.com

A cork sink made from Cork Polymer Composite. The detail shows the texture of the cork, bound in resin. As well as being aesthetically pleasing, the cork is very insulating, so a sink full of hot water would retain its heat longer than a ceramic sink. Images courtesy Simple Forms Design.

CASE STUDY

CORK TEXTILES

YEMI AWOSILE

Yemi Awosile—a textile and material designer, and an alumna of the Royal College of Art—became fascinated in the revival of manufacturing using cork due to its inherent qualities of flexibility, light weight, quietness and softness; traits "all useful", she states, "for a textile designer". In addition, the material benefits from a high level of elasticity, and has both sound and thermal insulating properties.

The re-composited base materials for both Awosile's Cork Fabric and chequered, laser-cut cork surface material are sourced directly from the wine industry. She is concerned with exploring the physical qualities of materials as well as their social impact. In this sense, the opportunity to use post-industrial cork waste was ideal. In addition, both applications can be used variously for upholstery, wall panels and wall coverings, and are available in numerous colours and metallic finishes. "I'm interested in colour", she writes, "so I work with print and weave techniques, combining cork with other materials to achieve an exciting visual aesthetic."

While researching, Awosile spent time in cork oak forests, or "Montados", in Portugal, learning about cork's production. Cork comes from the outer layer of the tree's bark and still has to be harvested by hand. She notes that once removed it reveals a beautiful orange coloured core. The cork trees in

LEFT
A sample of the tools and materials used to create the textiles.

RIGHT
Detail of Cork Textile 09.

TOP LEFT
Cork farming in Portugal.

TOP RIGHT
A cork farm in Portugal. The stripped trees are left alone for nine years to regrow.

RIGHT
Cork drying. Images courtesy Yemi Awosile.

Portugal are protected and never cut down. It is a really slow and yet efficient process. Once the bark is removed from the tree it takes nine years to regenerate, all the time absorbing CO_2 from the atmosphere. Cork, because it is farmed ecologically, without pesticides or irrigation, provides protective habitat for a large number of plants and animals.

She writes "I was impressed by the Portuguese industry and its producers because they found a way to nurture and industrially commercialise one of natures most interesting materials. A lot of the techniques used to harvest cork are ancient techniques which have stood the test of time."

BENDYWOOD

PROPERTIES
Can be bent by hand

APPLICATIONS
Furniture
Carpentry
Design

INFO
www.bendywood.info

Bendywood is a versatile hardwood moulding that can be bent from a cold, dry state, without the use of tools. It can then be used much like normal wood to create a range of design solutions. The benefits of using Bendywood are multiple—there is no need to use special tools as the wood bends to whatever shape you require; it can also be moulded back to a different shape if errors are encountered in its initial formation; and from a design perspective, the fact that the grain follows the curve of the wood means that it makes for a more aesthetically pleasing end-product.

Made from European hardwood beech, the wood is initially treated using a thermo-mechanical process to eventually become Bendywood. This process involves steaming the wood to soften the cells, and then, whilst damp, the wood is compressed and dried.

LEFT
Bendywood can be bent by hand to form tight bends without the need for steaming.

RIGHT
The handle for the spiral staircase at the Laban Centre, London, uses Bendywood for its long fluid spiral. Images courtesy Bendywood.

HONEYCOMB PAPER

Honeycomb Paper is as light as traditional paper, but has the added performance of being fire retardant. It is composed of a Polymer based fibre paper made of Aramid, Nomex and Kevlar, which is usually soaked in a Phenolic Resin. When stuck together in a honeycomb structure, a strong composite material is produced that can itself be used as a layer within other composite structures.

Variations of this kind of paper are used in aerofoil wings to give strength without adding weight, though it could also be used in any application where low weight to strength ratio and good resistance to heat are important factors. The aerospace industry uses it extensively. In fact, there are over 700 variations in shape and form, each with slightly different properties depending on the form the honeycomb paper is shaped into and the materials from which it is made.

As an example, Hexcel make a range of cell configurations in papers and metals. Expanding the hexagons in one direction gives greater flexibility, whilst reinforcing the hexagonal cells makes the overall form stiffer. Using different resins enables totally different applications, from underwater scenarios to situations of extreme heat or applications where huge amounts of flexibility are required.

PROPERTIES

Efficient energy absorber
Inherent dielectric strength
Mechanically tough
Flexible
Resistant

APPLICATIONS

Aerospace
Military

INFO

hexcel.com
professionalplastics.com

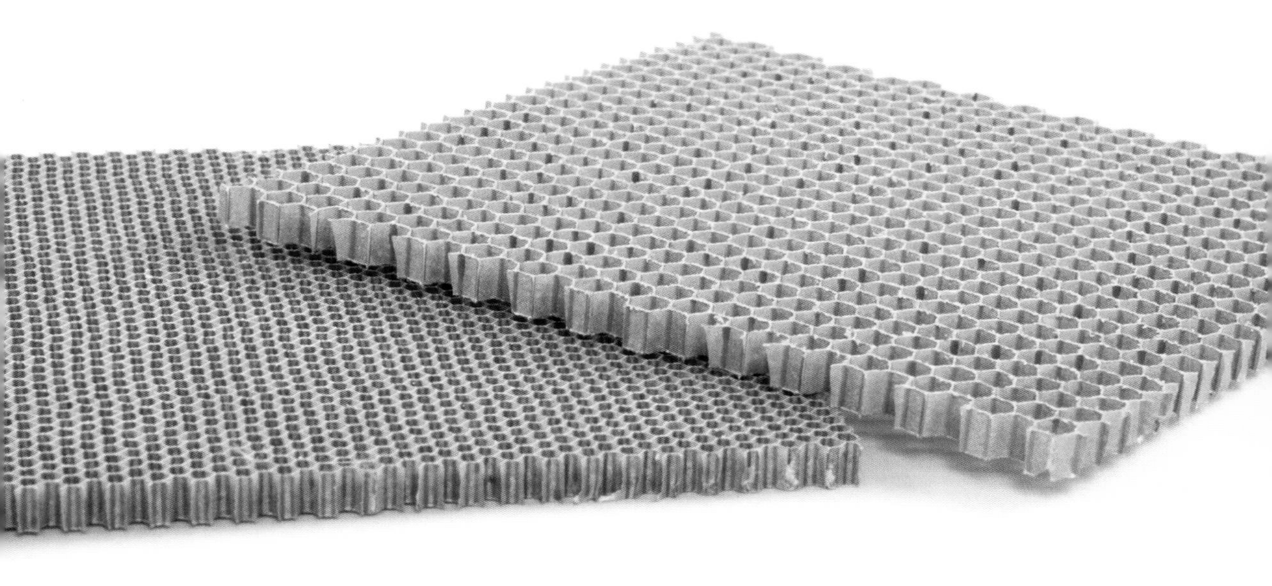

Samples of Honeycomb Paper made from aramid. The different forms lend the materials different structural properties.

LIQUID CRYSTAL POLYMER

PROPERTIES

Exceptionally inert

Fire retardant

Hydrolytic stability in boiling water is excellent

High chemical and

abrasion resistance

Resists stress/cracking in the presence of most chemicals at elevated temperatures

APPLICATIONS

Food containers and cookware

Liquid Crystal Polymer fibres

INFO

www.enricoazzimonti.it

www.ampulla.co.uk

www.sailingfast.co.uk

www.ticona.com/homepage

www.vectranfiber.com

Liquid Crystal Polymers (LCPs) are a class of Polyester Polymers. They are extremely inert, highly resistant to fire and often spun to produce fibres. Their properties are the result of their interesting molecular structure. Liquid Crystal Polymers have a more crystalline structure than most polymers. Instead of the molecular arrangement consisting of a mass of curved string of molecules, such as one might find in Nylon, the trands are aligned in an almost linear fashion, making the material anisotropic, lending it its strength.

Designer Enrico Azzimonti recently created a new form of cookware called Bloom using Liquid Crystal Polymer technopolymer material. The material is coated with a platonic silicone, making it non-stick. The clever design is multifunctional, able to be used as a shallow or deep pan or as a sealed pot. The use of LCP materials and silicone means that it is oven, microwave and freezer proof and as well as cooking in it, it is possible to preserve meals in it in the refrigerator or freezer. As the pans are made from plastics they are much less heavy than comparable metal or ceramic vessels.

Bloom, cookware by Enrico Azzimonti for Pavoni Italia. The range is made from an LCP Technopolymer, making it safe for the oven, microwave, refrigerator and freezer. Image courtesy Enrico Azzimonti.

LIQUID CRYSTAL POLYMER FIBRE

Vectran are worldwide leaders in the production of liquid crystal polymer (LCP) fibres. Vectran is a polyester-based high performance LCP produced by Ticona. It is naturally gold in colour and is similar to Kevlar 29, but has less strength loss with flex, making it great for applications where both strength and flexibility are key, such as for cruising sails where great force in difficult conditions and over long periods of time are required.

PROPERTIES

Gold in colour

Strong

Polyester-based

APPLICATIONS

Braided rope for high endurance

Medical applications

INFO

www.vectranfiber.com

WHITE BOARD PAINT

Idea Paint turns any surface that you can paint into a high-performance dry-erase surface—be it desks, doors, walls, hallways, or even old whiteboards. Durable and cost-effective, Idea Paint does not crack, peel or stain, and can be applied straight onto a surface with the use of a paintbrush and removed simply by painting over.

Idea Paint can be considered good for the environment as it can save on the use of paper in meetings and brainstorming sessions, though it unclear how much energy goes intothe making of the product. It claims to be the most environmentally responsible dry-erase product on the market, free from formaldehyde that is common in other such products and is known to decrease air quality in the immediate environment.

PROPERTIES

Durable

Cost-effective

Fast application

Environmentally friendly

APPLICATIONS

Whiteboards

INFO

www.ideapaint.com

Idea Paint can be used instead of paper for brainstorming sessions or group workshops. The writing is easily wiped away leaving only a blank wall. Image courtesy IdeaPaint.

Perspex® Fluorescent has a vivid, fluorescent edge which appears to glow under ambient light as though having its own light source. Images courtesy Perspex Distribution Ltd.

LIVE EDGE ACRYLIC GLASS

PROPERTIES

Can be easily laser cut

Can be worked with hand tools

Transparent

APPLICATIONS

A shatter-resistant alternative to glass

Retail display units

INFO

www.perspex.co.uk

Often referred to by its various brand names Perspex or Plexiglas, Polymethyl Methacrylate (PMMA), or Acrylic Glass, is a transparent thermoplastic often used as a shatter resistant alternative to glass.

Live-edge Acrylic Glass is just one example of a specific type of PMMA and derives its name from the way in which it interacts with light. Light which hits the plastic's surface is channeled through the plane of the sheet towards the edges, where a fluorescent glow is seen, making the edges appear to light up. It is a nice example of a relatively simple material that produces an unusual visual effect. Occasionally seen in retail display units, the material can be easily laser cut or worked with hand tools.

HYDROGELS

PROPERTIES

Can store large amounts of water

Hydrates

Non-toxic

Large volume expansion

APPLICATIONS

Use on or in the body

Gardening and forestry

INFO

www.sculpt.com

www.hydrogelvision.com

www.umcutrecht.nl

Hydrogels usually look like wobbly jelly. Structurally, they are comprised of an insoluble network of polymer chains that swell up when water is added, making an expanded mass.

There is a great variety of Hydrogels. They are often used as bandages for burns, and they are the material of soft contact lenses and implants. Made mostly of water, they are particularly effective for damaged skin because they are able to boost the self-cleaning action of wounds whilst keeping the area hydrated.

New innovations in the material are being developed all the time. Dutch researchers from the University Medical Centre in Utrecht have created Hydrogel 'ink' materials that can be 3-D printed to form stable cell-containing scaffolds that could potentially be used to graft body tissue or grow synthetic organs. Stem cells can be incorporated into the Hydrogels to kick-start the tissue regeneration within the scaffolds.

Researchers in Japan have made another breakthrough with Hydrogels, developing them to heal themselves. If cut with a razor the material will repair itself within three seconds. Self-healing Hydrogel is also tougher than most other Hydrogels, with solidity similar to silicone.

GLASS-LIKE RESIN

Still in the development stages but hopefully to be manufactured for market very soon, Glass-like Resin has an unusual property in that it can be reshaped and moulded even after it has been cured—a property which only certain inorganic compounds have possessed, until now. Repairable, recyclable and reversible, this unique material offers a number of opportunities for a range of industries including the aircraft and electronic industries, where resins, used for their lightness, strength and resilience, cannot currently be reshaped in the making of circuits and electronic devices. Another advantage is its low cost and ease of production.

Developed by a team of researchers led by Ludwik Leibler, CNRS researcher at the Laboratoire "Matière Molle et Chimie" (CNRS/ESPCI ParisTech), Glass-like Resin actually shares a number of properties of organic resins and rubbers, namely that it is light, insoluble and difficult to break.

PROPERTIES

Repairable
Recyclable
Reversible

APPLICATIONS

Automotive
Aeronautics
Car manufacturing
Building
Electronics
Sports industry

INFO

www2.cnrs.fr

LEFT
The material can take various forms.

RIGHT
A complex-shaped object is made by successively deforming and heating the material. Images © CNRS Photothèque / ESPCI / Cyril FRÉSILLON.

SPRAY-ON FABRIC

PROPERTIES

Fun application

Web-like

Hygienic

APPLICATIONS

Household goods

Fashion industry

Home decoration

INFO

www.fabricanltd.com

Fabrican is an instant non-woven fabric that can be used in a variety of applications, from household goods to fashion items, and, as the title suggests, it can be sprayed directly onto the body.

Voted "one of the 50 best inventions of 2010" by TIME magazine, Fabrican is applied with the use of an aerosol can or spray-gun and forms due to the cross-linking of fibres, which subsequently adhere to one another. In other words it can be considered as a spray-on felt-style material. Variation of colours, textures, and finished forms can be created using it due to its spray-can application, which enables easy coating, making it suitable for a range of different products and applications.

The wider application of Fabrican is also currently being considered, in particular with regards to large-scale projects such as household decoration and medical applications.

TOP
Fabrican can simply be sprayed out of a can directly onto the wearer's skin. Photograph © Gene Kiegel, courtesy Fabrican Ltd.

BOTTOM
A detail of Fabrican, the Spray-on Fabric, which can be stretched and will therefore move with the wearer's body. Photograph © LaytonThompson, courtesy Fabrican Ltd.

HEAT FUSIBLE YARN

Heat-Fusible Yarn is a polymer thread designed to melt at a low temperature, usually between 70 and 140 degrees Celsius, or 158 and 284 degrees Fahrenheit, and is frequently used in the textile industry to fuse fibres together. Time, and therefore money, is saved when using this material as it can create hems at high speed and achieve very precise and quick joins.

Ordinary yarns are made up of continuous strands of twisted threads, of natural or synthetic materials, but their main flaw is that stray fibres are easily pulled out of the basic fibre pile. By integrating heat-fusible yarns made from low-melt co-polyamide or co-polyester, a fabric can be made stronger with heat, as the integrated yarn fibres melt to form an adhesive.

Designers Shota Aoyagi & Jungeun Lee created a series of works called *Nuue* using Heat-fusible Yarn as a sculptural material. As well as creating beautiful and novel forms of garments, the designers have said that the process has potential for product designers, both for trialling soft forms and quickly and inexpensively building shapes.

PROPERTIES

Can bond fabric using only steam

Can be sewn into regular fabrics

Low melting temperature 70–140 degrees Celsius

APPLICATIONS

Chenille and fantasy yarns

Embroidery

Garments and tailoring

Iron-on fabric

Building soft forms

Sculpture

INFO

www.swicofil.com

www.harodite.com

www.districo.com

www.studiokoya.com

LEFT
The yarn is wound around a form and steamed. The heat from the steam causes the fibre to retain its shape and bond to other fibres.

RIGHT
A garment in the *Nuue* collection. The fabric has kept the form created from its making, therefore creating a garment capable of holding its shape. Images courtesy Studio Koya.

SPACER TEXTILES

Spacer Textiles are a unique type of 3-D textile designed to have inbuilt give and cushion qualities, whilst allowing air to circulate through the structure, preventing build-up of moisture. Usually consisting of two layers of fabric separated by yarns that are connected at a 90 degree angle to the layers, the cushioning abilities of the material can be engineered to suit a variety of needs.

This prototype chair, *Spacer Chair* by Droog Design and NEXT Architects, is constructed using fibreglass with Nylon spacer fabric and polyester resin. The chair pushes the technology of spacer fabrics to a new level, expanding their scale and using their inherent bounciness as part of the design. By using a double weaving technique commonly found in carpets, where the plush threads visible on the surface are the result of two carpets being woven together and cut loose, the chair is given strength by the double thickness and also the use of curves. The resin gives the final structure its stiffness. The project demonstrates that when scaled up, Spacer Textiles' sculptural forms come into their own, but also that their stiffness and inherent structure could have a number of interesting uses in larger scale design projects.

PROPERTIES

3-D fabric structure

Cushioning

Good air circulation

Variable stiffness

APPLICATIONS

Carpets

Furniture

Potential use in larger scale design projects

INFO

www.droog.nl

www.nextarchitects.com

www.muellertextil.de

www.terrot.de

Spacer Chair by NEXT Architects & Samira Boon. The Spacer Textile has been made stiff and able to support weight by being impregnated with resin. Images courtesy NEXT Architects.

INSTANTLY-HARDENING FABRIC

PROPERTIES

Instantly hardening
Shock absorbing
Breathable
Durable
Washable

APPLICATIONS

Body armour
Electronics protection
Sportswear

INFO

www.dowcorning.com

Dow Corning, a global leader in silicones, has recently created a fabric which instantly hardens when impacted. Made from polymerised silicone sheets, the fabric called Deflexion, has a molecular arrangement which changes, tightening together around the area of impact. This is what causes the material to instantly harden and spreads the impact across the material.

It is soft, flexible and breathable, making it comfortable to wear while doing strenuous activity. It has recently been trialled for sailing clothing for use in harsh weather because, being made from silicone, it can easily get wet.

TOP
The TP-range, which is not flexible but is available in very thin sheets, at a minimum thickness of three millimetres.

BOTTOM
The S-Range, demonstrating this material's flexibility. The fabric is available in three different thicknesses.

Renderings demonstrating how the material might be used as impact protection for sports or body armour applications. If impacted, Dow Corning's S-Range, which is more flexible and breathable than the TP-Range, will harden, spreading the force of the impact and therefore protect the wearer. Images courtesy Dow Corning.

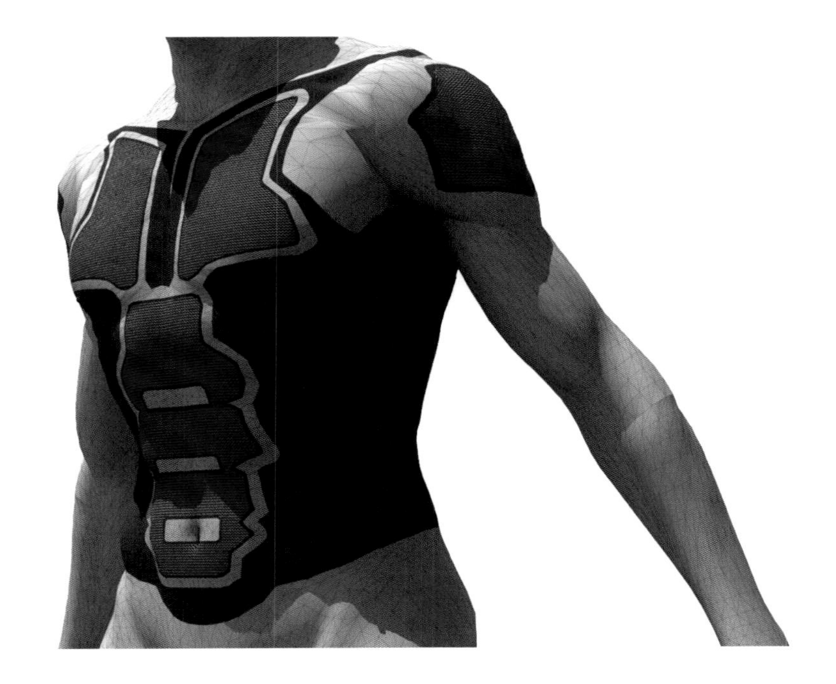

POLYCAPROLACTONE

PROPERTIES

Very short degradation time
Completely biodegradable
Strong
Low working temperature
Solid at room temperature

APPLICATIONS

Adhesives
Prototyping
Additive in other thermoplastics
Sculpture

INFO

www.InstaMorph.com
www.inventables.com
www.jerszyseymour.com
www.perstorp.com

Although Polycaprolactone isn't produced from renewable materials—rather crude oil—it is completely biodegradable. In industry it is used in the process of making other products, such as synthetic leather or stiffeners or adhesives. Without the addition of other materials it can be processed into biodegradable plastic bags, though its quick decomposition makes it more fragile than other alternatives.

As a material it is possible to buy it as a wax which, when heated, becomes workable at around 60 degrees Celsius or 140 degress Fahrenheit. At room temperature it is a solid mass with quite high strength. Designer Jerszy Seymour ran a series of installations with the wax, creating quick joints for creating objects in a relatively short space of time. *Workshop Chair* is one such outcome and demonstrates the strength of Polycaprolactone. In independent testing the chair was able to withstand 200 kilograms or 440 pounds of pressure on the seat.

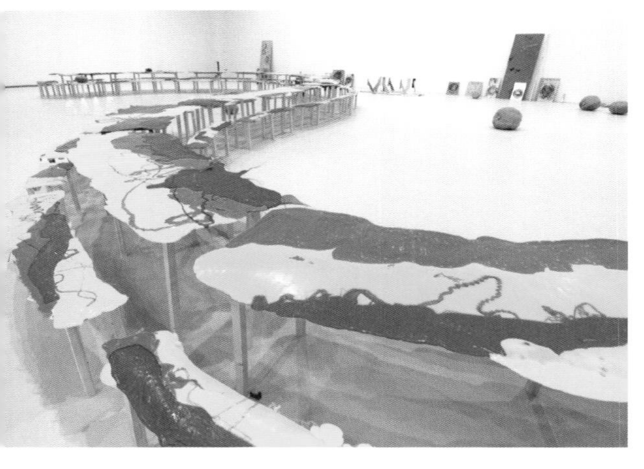

LEFT
The first supper that took place at the MAK in Vienna. The table and long benches were made by pouring brightly coloured Polycapralactone wax on the floor.

RIGHT
An arrangement showing *The Workshop Chair*. This chair's joints are constructed from Polycapralactone wax, which, when cooled, creates a solid joint, capable of holding over 200 kilograms or 440 pounds. The 'Amateur Workshop' was first shown at galeries des Galeries Lafayette in Paris 2010. *The Workshop Chairs* were first developed for the exhibition 'Coalition of Amateurs' at the Mudam in Luxembourg in 2009. Images © Peter Kainz and courtesy Jerszy Seymour Design Workshop"(JSDW).

PIEZOELECTRIC PLASTICS

Piezoelectric Plastics are plastics that produce an electric charge once activated. The activation can take many different forms, such as pressure or pulling. The materials can be processed using techniques usual to the manufacture of ordinary plastic, though the thinner they are produced the more likely a small activation has of producing a reaction. A commonly used Piezoelectric Plastic is Polyvinylidene Fluoride (PVDF). They are widely used in areas such as surveillance where pressure of some sort is needed to activate an alarm. Piezoelectric Polymer films have also been developed which can be used in loudspeakers and as large-scale sensors.

Recently a number of design and engineering projects have explored the possibility of generating electricity using PVDF to contribute to the power grid. The *New Scientist* reported that Jean-Jaques Chaillot at the atomic energy commission (CEA), Grenoble, France, had developed a way to generate electricity from rainfall. Still in its infancy, the project could potentially be used to supplement the power generated by solar panels at night or on dark and rainy days.

PROPERTIES

Conducts electricity

APPLICATIONS

Architecture

Speaker systems

Surveillance systems

Medical diagnostics

INFO

www.oceanpowertechnologies.com

www.cea.fr

ELECTROACTIVE POLYMERS

These are a class of smart materials that change shape when an electric charge is applied. They do this by converting the electrical force into kinetic force, but they also function the other way around, converting kinetic energy to electrical energy, which means that they are the subject of much research for renewable power generation. There are two common forms: dielectric and ionic.

NASA are conducting research into these materials for use in mechanical robot arms and to make aircraft manoeuvre in the air more like birds than conventional planes, not mimicking the flapping action but rather flexing the wing to adjust airflow underneath and therefore turning without the need of so many mechanical parts. The materials are also currently being considered for building skins, responsive structures and prosthetics. Designer Stefan Ulrich has worked with FESTO and EMPA to create a pillow that mimics breathing called Funktionide. The idea is that the motion of a chest moving with steady breaths creates a comforting sensation.

PROPERTIES

Flexible

Change shape when electrical current added

APPLICATIONS

Prosthetics

Aircraft

Architecture

Responsive surfaces

INFO

eap.jpl.nasa.gov

www.eltopo.de

CASE STUDY

SHAPE-CHANGING ARCHITECTURE

SHAPESHIFT

Using Electroactive Polymers, a group of students at the ETH in Zurich have collaborated with the Swiss Federal Laboratories for Materials Science and Technology to develop a project called ShapeShift. The project explores the potential of non-mechanical kinetic architectural materials.

Each individual element is made from a very thin layer of stretched acrylic tape coated with conductive carbon black powder that is then insulated with a fine layer of liquid silicone. A charge moves the film in two directions: squeezing it in and elongating it. As each film is pre-stretched when it is attached to flexible frames, the whole form is bent, which means that when the charge is applied, the form relaxes and straightens. The curve of individual forms can therefore be altered with the application of electrical current.

The team experimented with a number of variations in the component arrangement, first using individual panels attached to a larger frame, but eventually their research led them to create forms comprised of direct component-to-component linkages networked together to produce self-supporting forms. These self-supporting forms have the potential to become dynamic shifting architectural systems and responsive environments capable of adapting to human input and external environments.

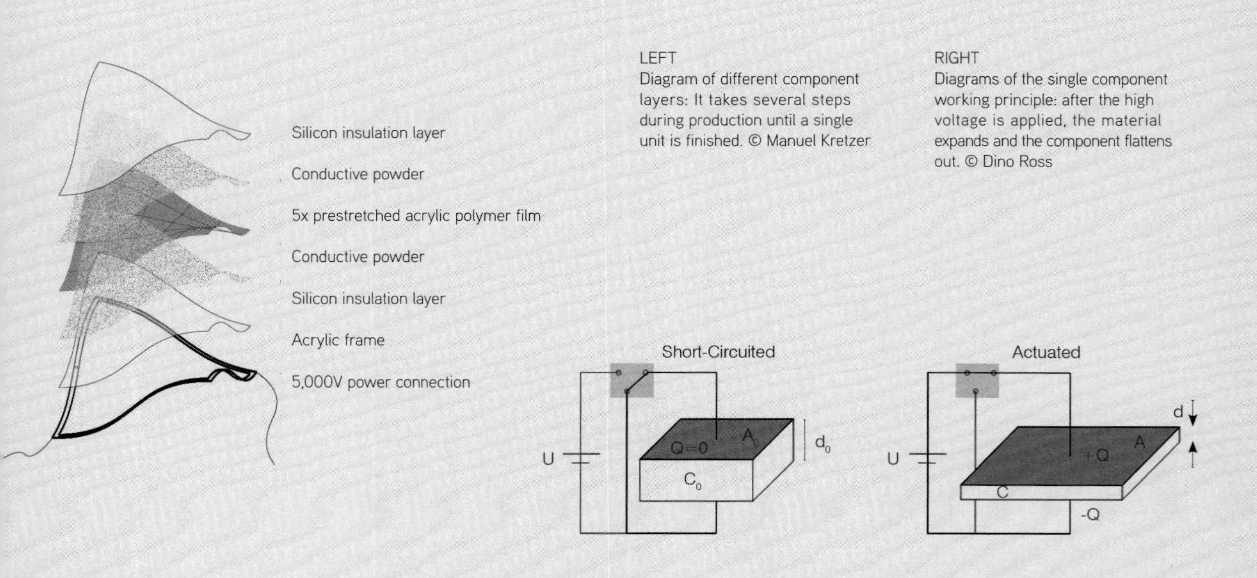

LEFT
Diagram of different component layers: It takes several steps during production until a single unit is finished. © Manuel Kretzer

RIGHT
Diagrams of the single component working principle: after the high voltage is applied, the material expands and the component flattens out. © Dino Ross

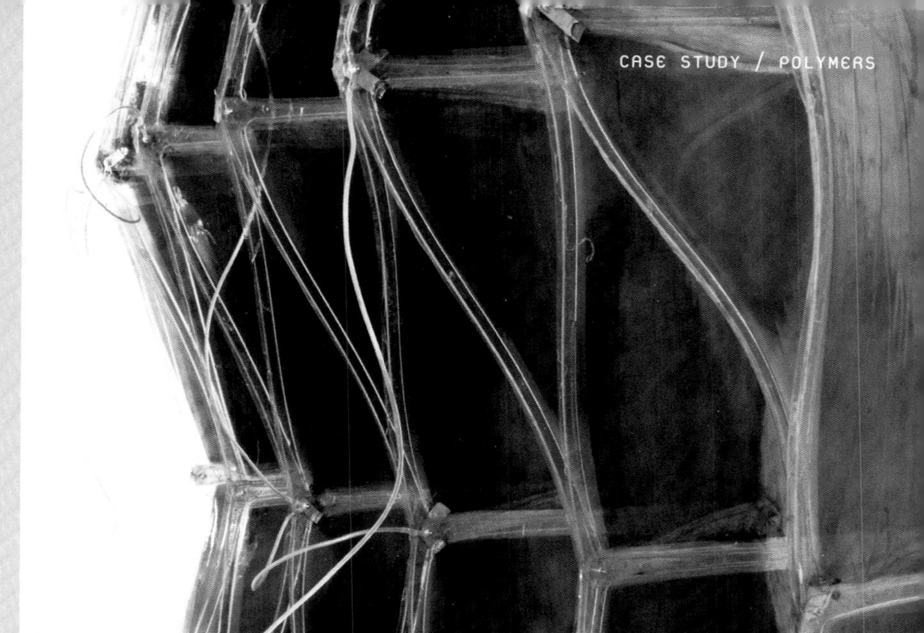

TOP
Details of final structural arrangement: all components are connected to each other to produce a self-supporting form. Photograph © Edyta Augusynowicz

BOTTOM
ShapeShift final installation. Exhibited at Starkart Gallery, Zurich. © Edyta Augustynowicz, Sofia Georgakopulou, Dino Rossi and Stefanie Sixt, CAAD ETH Zurich. Photograph © Manuel Kretzer.

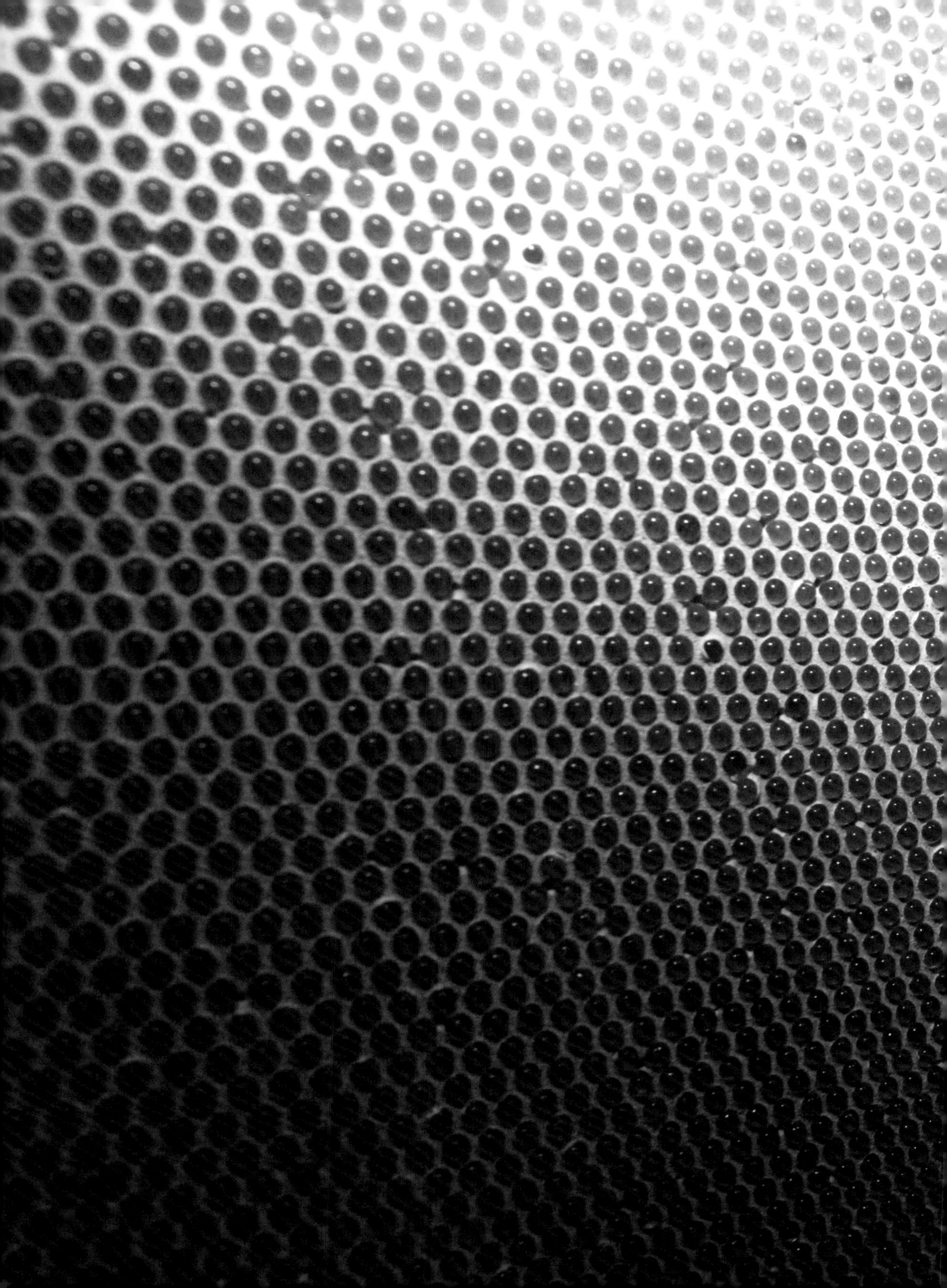

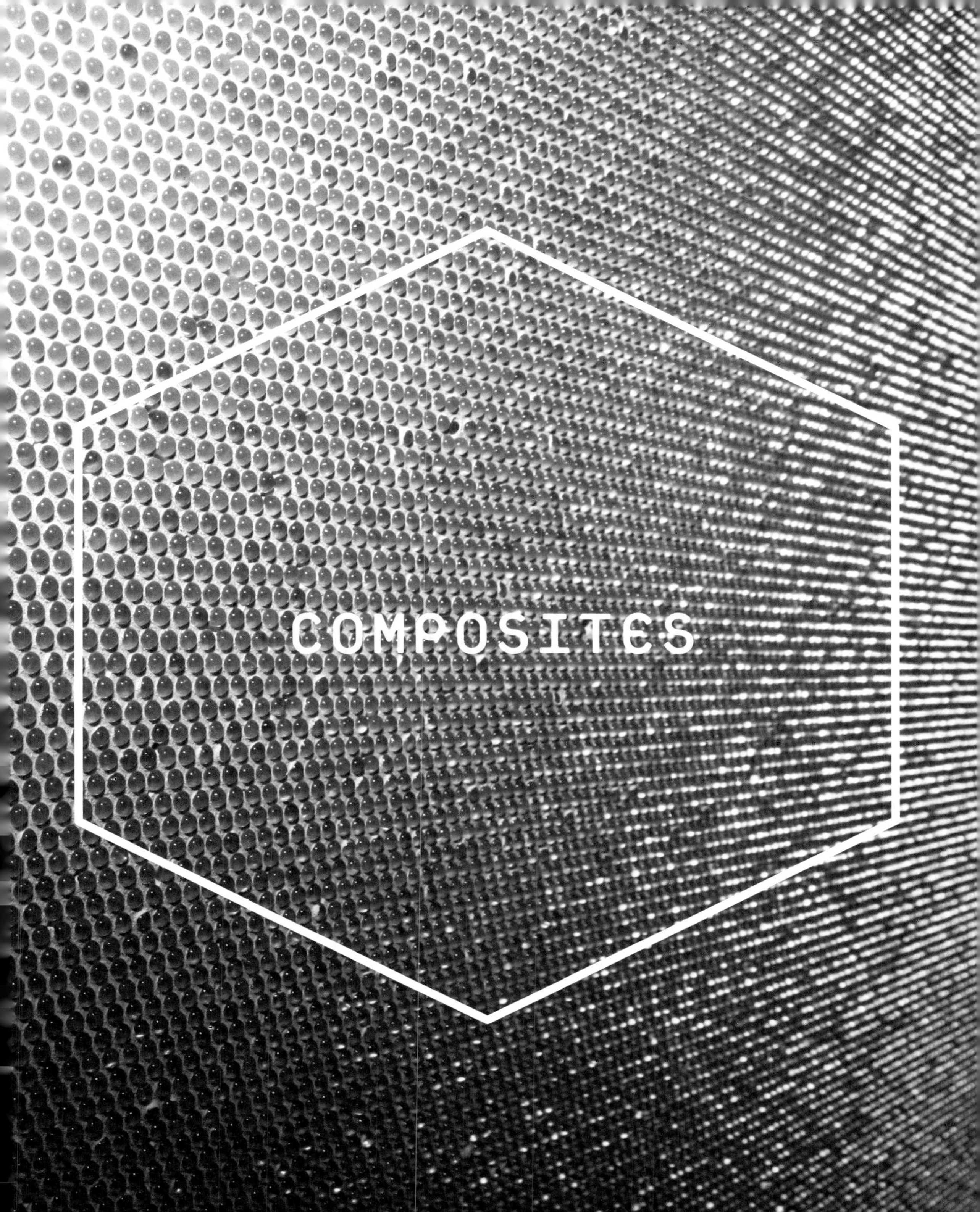
COMPOSITES

The creation of composites makes novel materials that are better than the sum of their parts, allowing us to push materials beyond their compositional limits. They are made by combining two or more materials with significantly different properties to yield a new material with a character distinct from those of its constituents. These components do not blend or dissolve into one another; instead they are assembled into a distinct structure that gives rise to cooperative and constructive behaviour. Common examples of composites include concrete (cement combined with gravel), plywood (layers of differing types or orientations of wood) and Fibreglass (glass strands suspended within a polymer binding). In each of these examples, the final material exhibits greater strength and resilience to certain forces than each of the constituent parts alone would, but are still distinguishable in the mix at a macro level—that is, we can still see them with the naked eye.

In recent years, composites have been the focus of much material science and engineering for applications in high-performance arenas such as aerospace and Formula 1 motor racing. The most famous example is that of carbon-fibre-reinforced-polymer, where long black Carbon Fibres are woven together and embedded in thermoset resins. Carbon fibre composites are the strongest, stiffest and lightest engineering materials currently in existence. They have become so readily associated with their superior material performance that the characteristic visual effect of this material is occasionally copied as a surface finish, applied to standard polymers, to trade on the connotations of the material's aesthetic.

Not all composites are high-tech, however. The idea of combining different materials together to achieve performance beyond that of the singular material is nothing new. Wattle and daub, for example, is a traditional composite building material that has been used for thousands of years to create homes for civilisations across the globe. The key compositional elements are straw and a wet binder that will dry hard. This was often mud, clay, sand and even animal dung. Another example is waxed cotton, which is made by impregnating woven fabric with paraffin wax. The resultant cloth exhibits waterproof properties that are still exploited to this day in some jackets and sail cloths.

Composites that occur in Nature serve as a prime example of how materials with miraculous characteristics can be formed by combining a few simple constituents. Biology only uses a small number of biopolymers, a few minerals and some cross-linking agents in its construction endeavours. However, a huge variety of incredible composites arise in the natural world, from hard teeth and bone to strong and durable trees to elastic but tough skin. These properties are achieved through hierarchical assembly of structure, rather than through chemical composition alone. A good example is nacre ('mother of pearl'), the tough ceramic inner layer of mollusc shells. It is over 99 per cent calcium carbonate (the same stuff as chalk), with a small amount of proteins embedded. Tiny plates of calcium carbonate are bound together by a proteinaceous glue, like bricks-and-mortar, a structure which makes nacre thousands of times tougher than the calcium carbonate alone. In normal ceramics, cracks can propagate rapidly through the structure with little resistance. In nacre, the cracks are stopped from spreading by the protein interfaces, which absorb and dissipate the energy.

PH and ZL

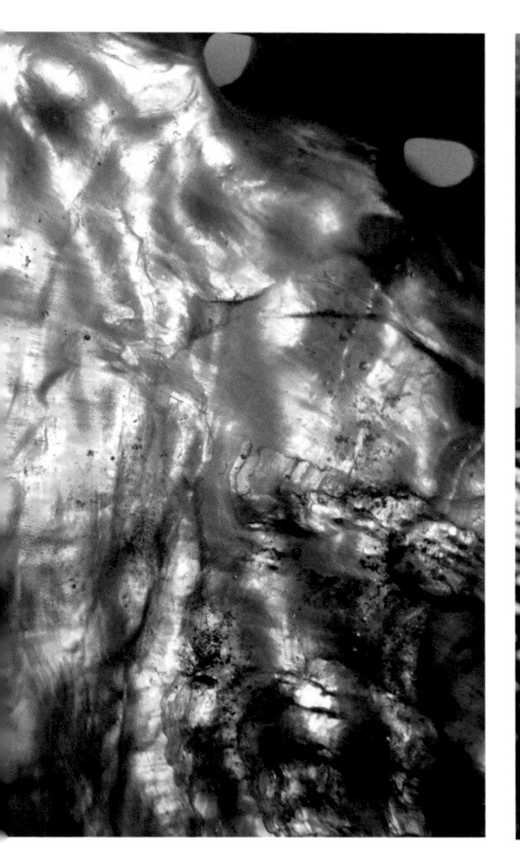

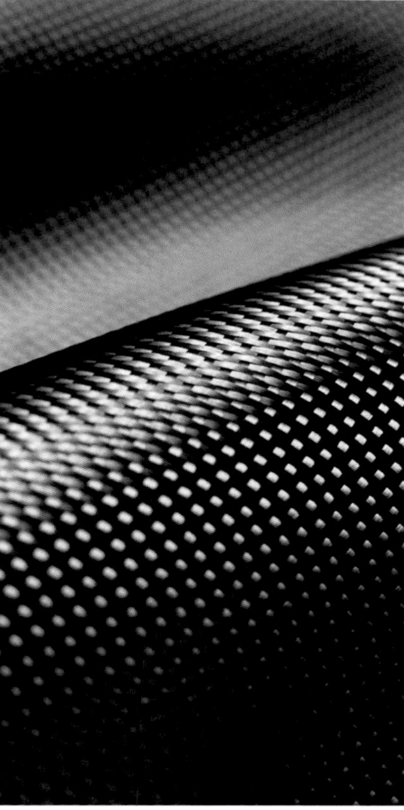

PREVIOUS PAGES
A detail of Blingcrete being activated by light. Image courtesy Blingcrete™.

LEFT
A detail of nacre, one of nature's composite materials.

RIGHT
A detail of carbon-fibre-reinforced-polymer, the material of choice for racing cars, aerospace engineering and sporting equipment.

LIGHT-REFLECTING CONCRETE

PROPERTIES

Multiple variations and textures

Structural

Decorative

Creative freedom in surface layout

APPLICATIONS

Signs

Safety measures

Architecture

INFO

www.blingcrete.com

www.atlerierk10.de

With the name BlingCrete™, it is of no surprise that this innovative new product is an ultra shiny Concrete. Using only low-tech modes of reflection, the concrete is able to reflect vast amounts of light. It works in the same way that road paint reflect headlights back brightly through the inclusion of millions of tiny glass spheres embedded in the Concrete's surface.

BlingCrete™ use new high-performance Concretes, adhesives and casting techniques that enable them to control precisely the position of the glass beads in the matrix of the material, in turn controlling the level and pattern of the reflected light. Although it has huge decorative potential, other possible applications include embedded safety marking along the edges of danger areas such as steps, curbs and railway platforms. Because it simply reflects light, its glow can be altered with the introduction of external light sources from different angles. Different sizes of glass bead can also be embedded to create a variety of effects.

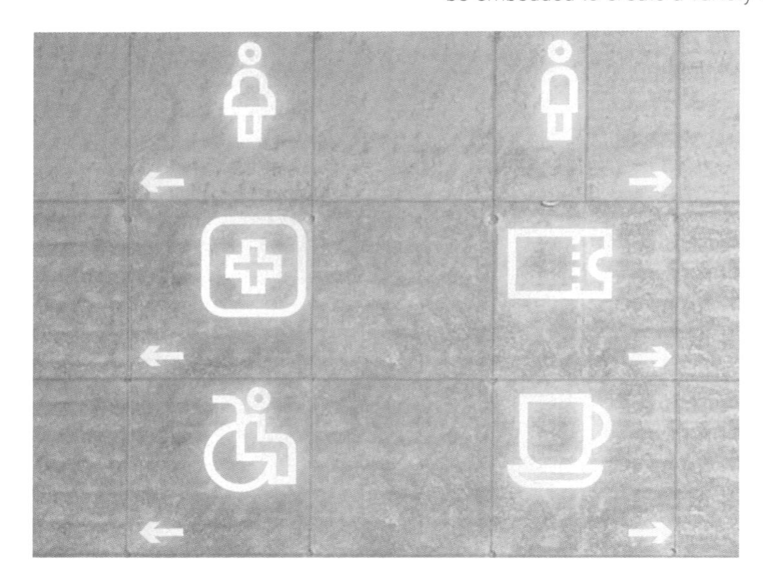

TOP LEFT
Using Blingcrete it is possible to create signs which are only visible when activated by lights, such as the headlights of cars or another directional light.

LEFT
Close-up view of concave mock-up demonstrating the different lighting effect achieved with different sized glass spheres.

TOP RIGHT AND CENTRE
Images showing Blingcrete in light above and darkness with only directional light activating the material below. Images courtesy Blingcrete™.

TRANSPARENT CONCRETE

Concrete, which is typically solid grey and opaque, can be turned transparent by the inclusion of Glass Fibres. Along with their ability to carry light, these fibres have the added benefit of structural integrity, which should allow the manufacture of strong Composite Concretes. If the Glass Fibres are aligned in close packed matrix, it will be possible for light to pass through in significant amounts.

Litracon™ manufacture such a Transparent Concrete as a building product in which they combine Glass Optical Fibres and fine Concrete. It can be produced as prefabricated building blocks and panels. Due to the small size of the fibres, they blend into the Concrete, becoming an integral component of the material.

Strong and transparent, Litracon's current Concretes are simultaneously structural and decorative building materials, but their ability to appear transparent against light and opaque against darkness lends itself to other applications, from security screens to spaces where windows are not permitted.

PROPERTIES

Transmits light

Strong

Structural

APPLICATIONS

Architecture

Decorative

INFO

www.litracon.hu

Litracon is installed in the entrance of the Museum Cella Sephichora, Pecs, Hungary. From inside the building it is possible to see the silhouettes of people in the street. Image courtesy Litracon.

SUPER-STRENGTH CONCRETE

PROPERTIES

High packing density

High compression stability

High resilience

Low cost

APPLICATIONS

Decorative purposes

Interior design

Architecture

Product design

INFO

www.gtecz.com

www.timmackerodt.de

Super-strength Concrete, a reformulation of traditional concrete, was developed using advanced engineering and mathematical principles which gave rise to a material that stretches the boundaries of what Concrete is capable of. It is therefore used in the expected applications of buildings and infrastructures but designer Tim Mackrodt has created a line of domestic products with it (a lamp and a table series), which show-off the material's capabilities. The series is cast to have fine texture and thin filigree walls of exceptional strength.

LEFT
Detail of Tim Mackrodt's *FALT.stool*. The stool saves weight and gains stability from an undercut structure underneath the seating surface.

ABOVE
FALT.series Table and *Lamp*. The concrete is supplied by G.tecz from Kassel, but Mackrodt works with it, folding the concrete in thin layers to create a unique surface. The concrete lamp only weighs 1.4 grams and has a wall thickness of only 2.7 milimetres. Images courtesy Tim Mackrodt.

GRAPHIC PATTERNED CONCRETE

One of the recent innovations in the world of concrete comes from a Finnish company called Graphic Concrete Ltd, who have developed a process of creating graphic patterns on the surface of concrete.

To create Graphic Concrete™, one applies a special surface retardant coating to a membrane in the pattern required. This membrane is then placed in the bottom of the mould into which a concrete slab is cast. Once the concrete is set, the slab is turned out and the membrane if washed away to reveal the required pattern. The flexibility of this method allows the creation of slabs of varying dimensions carrying an image that can be printed in a binary design.

To date, the manufacturers have worked with a number of designers to create images of varying complexity on walls, floors and feature panels in both internal and external environments and the technique is becoming increasingly recognised as a quick and simple way to render and embed texture and graphic detail on the surface of concrete.

PROPERTIES

Creates patterns

Flexible

Surface texture

APPLICATIONS

Architecture

INFO

www.graphicconcrete.com

ABOVE
Details of the pattern on the walls of the Skanska Headquarters.

RIGHT
The pattern in the concrete continues onto the window surround. Graphic Concrete can be used for branding on the buildings. Images courtesy Graphic Concrete.

CREACRETE

PROPERTIES

Abrasion Resistant

Glossy

Acid resistant

Food safe

Hydrophobic

APPLICATIONS

Alternate material to ceramic

Interior design

Architecture

INFO

www.alexalixfeld.com

Developed by designer Alexa Lixfeld, Creacrete™ is a concrete based material, which is highly dense and compact, making it possible to create filigree and thin-walled objects from concrete. In effect, what would previously be made from ceramic materials can now be reproduced in concrete, without the need for firing. Special processing makes it possible to achieve a high gloss surface that is new to concrete. A nanoscale engineered coating makes cups and plates produced in it hydrophobic and food safe.

Creacrete is an alternative to ceramics for floor and wall coverings, decorative objects and building facades. Because it is made in a cold-casting process it reduces costs and energy consumption. Exploring the aesthetics of concrete while introducing new traits, Creacrete promises surfaces that are permanently glossy, food-safe, abrasion and acid resistant and thus a myriad of new applications.

ABOVE
A floor made from Creacrete would be very easy to clean and would be a good material for high traffic areas such as a commercial kitchen. It could also be used for countertops because it is food safe.

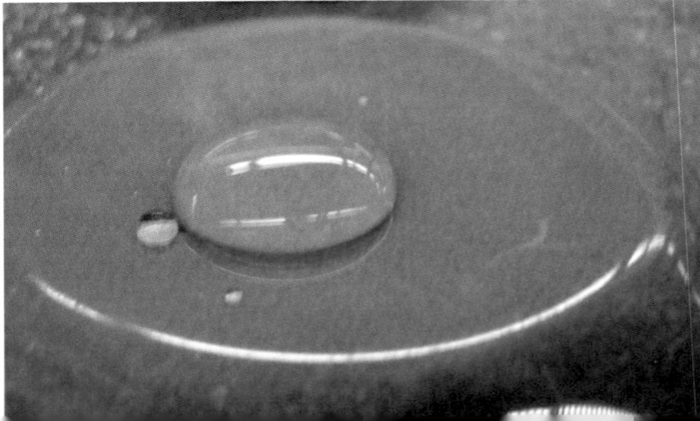

RIGHT
A sample of Creacrete demonstrating its extreme hydrophobic qualities. The droplet of water will not sink into the surface and therefore remains held by its own surface tension. Images courtesy Alexa Lixfeld.

3-D CONCRETE TILES

These decorative 3-D Concrete Tiles come in a range of patterns from Metrofarm, a design company based in Berlin. Each design has an individual, raised pattern and is constructed differently in order to produce a fresh and self-contained visual idea. The Citroen shaped arrows and swirls are formed in concrete moulds, wet concrete poured into set concrete. Used for indoor applications, each tile is delicately hand-made, lending the very industrial material a new, softer and more decorative aesthetic.

Julia Böttcher and Jan Müller founded Metrofarm in 2001. They accept international assignments and their speciality practices are design and manufacture, from preliminary planning through to the finished product and installation.

PROPERTIES

Smooth

Attractive

Graphic

APPLICATIONS

Interior design

INFO

www.metrofarm.net

The Sahnehaube and Citroen tile patterns, 2007. Image courtesy Metrofarm.

PIGMENTED CONCRETE

PROPERTIES

Coloured

Does not require additional surface treatment

Colour unaffected by chipping

APPLICATIONS

Architecture

Furniture

Sculpture

INFO

bayferrox.com

As a building material, concrete is often defined by its grey colour. However, there are various products available to pigment concrete, allowing for the creation of a structural material with embedded inherent colour. This means that when a specific colour is called for, a concrete can be cast that does not require additional painting or surface treatment and if it were to chip, the colour beneath would be consistent with the surface of the material.

A number of manufacturers make inorganic pigments that offer highly consistent colours when combined with concrete, though the pallet of possible colours is limited.

STONE-LIKE TILES

PROPERTIES

Recycled/recyclable

Sturdy

Traffic and weather resistant

APPLICATIONS

Indoor and outdoor tiling

Architecture

INFO

www.angelograssi.it

The aesthetic of natural stone is still much admired for tiling in interior and exterior applications, but it is an expensive option. A recent solution comes in the form of tiles that mimic the appearance of natural stone but are in fact composed completely of recycled materials, and are themselves completely recyclable.

More than half the material composition of Ripietra tiles, made by Angelo Grassi Design, is made up of Polyethylene found in urban waste. Another 45 per cent is composed of wood from industrial processing waste. It is a Composite material produced by extrusion and moulding to take the form of stone slabs. The tiles are suited to both interior and exterior environments, they are anti-slip, and can withstand the weight of road vehicles. Although not natural stone, the material is strong, weather resistant and does not lose its shape in direct sunlight.

Ripetra tiles are formed into stone shapes and laid on grass to demonstrate their use outdoors. The material does not fade in sunlight. Image courtesy Ripietra.

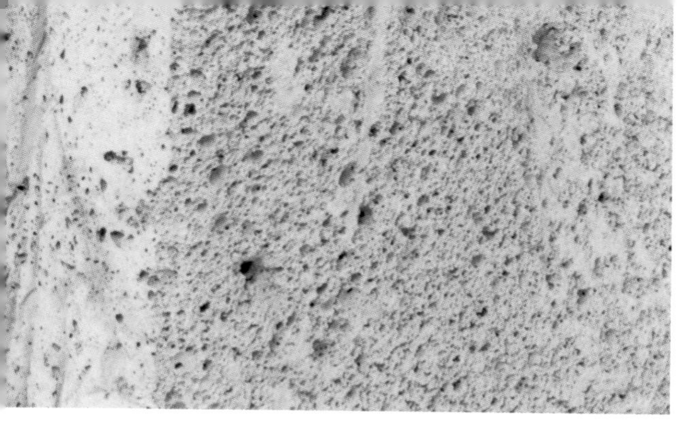

Detail of the texture of the foam. The cross-section allows the bubbles to be seen in the structure.

CONCRETE FOAM

Concrete Foam uses a foaming agent and foam generator to aerate preprepared concrete slurry, making it much lighter—a bonus in the building of quick and prefabricated housing. Concrete Foam is also really useful for patching cracked existing concrete and infilling holes.

Foaming agents are classified by the basis of the active ingredient. The earliest varieties were protein based, using proteins from animal carcasses, but this lead to variations in the quality and a terrible smell. Surfactant based (synthetic foaming agents) can also be problematic, with inefficient manufacturing processes, though a more consistent quality. Enzyme based foaming agents are a new and innovative foaming agent technology used in LithoFoam by Dr Lucà & Partner. LithoFoam consists of highly active proteins of mainly vegetable origin.

PROPERTIES

Available in a variety of weights and densities

Strong

Reduces the gross density of concrete from 2400 kg/m³ as low as 100 kg/m³

APPLICATIONS

Architecture

Repairs

INFO

www.dr-luca.com

RICE CEMENT

Peoples around the world have historically developed their building techniques using the materials closest to hand and most easily produced. For example, in ancient China, sticky rice was added to construction mortar to form a limestone cement, stronger and more water-resistant than its non-rice filled rival.

Miniwiz, a design, engineering and manufacturing company, have revived and renewed the technique, but use agricultural waste instead of newly grown rice products. Their process extracts silica from barley and rice husk, creating an agro-silica—a nanoscale additive capable of improving strength, workability and fire proofing. At the same time it cuts out the toxic chemical enhancers such as silica-additives so often included in high performance cements.

Testing suggests that the Rice Cement could reduce concrete costs by 15 per cent because of its improved compressive strength, allowing for less to be used. Rice Cement is has very low porosity which prevents mould growth, chemical erosion and other damage, and therefore improves its lifespan.

PROPERTIES

SiO_2 content is over 98 per cent

Better lifespan than non rice-filled concrete

APPLICATIONS

Architecture

Furniture

Engineering

INFO

www.miniwiz.com

www.ricecement.com

CONCRETE SHELTERS

CONCRETE CANVAS

Founded by Peter Brewin and Will Crawford, Concrete Canvas was originally conceived as a building in a bag that requires only water and air for construction. Like many great innovations involving materials, it has since developed to have a number of varied applications, from drain linings to hillside protection, but the application with the most immediately life-improving results is undoubtedly the Concrete Canvas Shelter—their first project concept, which they have now refined and used to great effect in humanitarian crises around the world.

Concrete Canvas is made from flexible concrete impregnated fabric that hardens when hydrated and dried. The composite is constructed from a 3-D synthetic fibre matrix with a PVC coating on the interior surface. The fibre acts as a wick, drawing water into the impregnated cement. Once hydrated, the fabric has a two hour working time until it hardens. The company website states that "a CCS25 variant (Concrete Canvas Shelter 25) can be deployed by two people without any training in under an hour and is ready to use in only 24 hours. Essentially, Concrete Canvas Shelters are inflatable concrete buildings." The material is so simple to use it can even be cut using basic hand tools prior to setting, and once set, can be ground with an angle grinder. It can even be hydrated using seawater.

OPPOSITE
A Concrete Canvas Shelter deployed and secured with a door, an ideal set-up for military use.

ABOVE
Images of Concrete Canvas Shelter's deployment in the field. The inflatable core can be operated by two people and folds up small enough to fit on the back of a truck.

RIGHT
The shelter, once hardened, can be cut and modified with an angle grinder. The image shows how thin the layer of Concrete Canvas is. Even at this thickness the form is strong and can be secured. Images courtesy Concrete Canvas.

The shelter is deployed by inflating an interior structure which supports the hardening concrete canvas shell. Once set, the inflated centre can be removed leaving a hard solid shell that can be secured with a door—a huge advantage when in difficult areas where security is important and looting a potential threat. The interior can even be sealed to create a space where life-saving operations can securely take place without risk of infection.

Research was conducted into the material in a variety of extreme survival situations ranging from Uganda to New Orleans in the period after Hurricane Katrina. The product was tested in the field, interviews were undertaken with twenty two UN Agencies and NGOs and a feedback has been received from a forum in Geneva with staff from the all the major aid agencies, with the goal of making this material as good as it can be at meeting the needs of people in need of emergency shelter.

SELF-HEALING CONCRETE

PROPERTIES

Can reduce production of CO_2

Lengthens life of concrete

Potentially resistant to corrosion

APPLICATIONS

Architecture

Engineering

INFO

www.forbes.com

www.uri.edu

Self-healing Concrete is a smart material; it reacts to environmental triggers and heals itself when stressed. Regular concretes contain calcium hydroxide, but a recent development in Self-healing Concrete contain a healing agent, sodium silicate, which reacts with the calcium hydroxide when cracked or damaged. This creates a gel-like material that hardens in about a week, blocking the pores in the concrete and re-strengthening the weakened material.

There are a number of Self-healing Concretes involving filling the concrete with bacteria spores that secrete calcium carbonate, or with tiny glass capillaries containing a healing agent, but these methods are costly. The use of sodium silicate is likely to be cheaper and produce concrete that rivals conventional concrete in cost. It was developed by a student at the University of Rhode Island, Michelle Pelletier, who is now working on a study to determine whether the sodium silicate healing agent is also able to act as a corrosion inhibitor, preventing the common problem of corrosion within the steel reinforcements embedded in concrete.

Her discovery might lead to a concrete able to reduce current CO_2 emissions from the cement industry by 20 per cent and significantly lengthen the life of built structures.

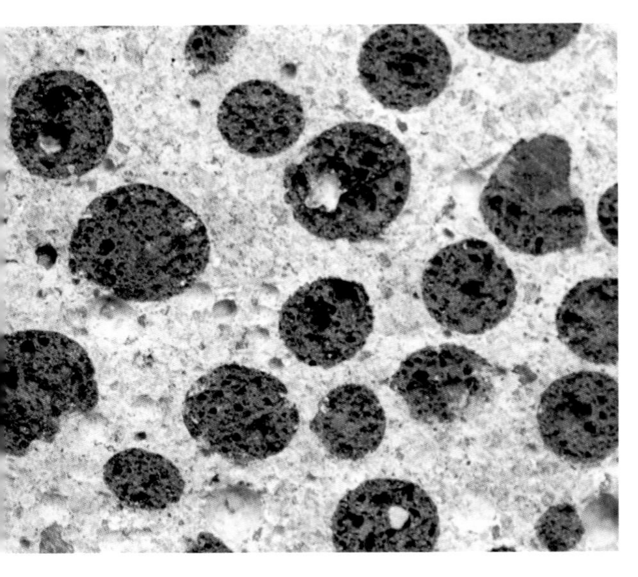

LEFT
A detail of the surface of Self-healing Concrete that uses bacteria spores to mend cracks.

RIGHT
A sample of Michelle Pelletier's Self-healing Concrete. Image courtesy University of Rhode Island.

BRICKS MADE FROM WASTE PRODUCTS

University of Leeds spin out, Encos Limited, has been working with Yorkshire Water on an unusual project with the potential to drastically reduce the carbon footprint of the construction industry.

By curing the ash from incinerators including the likes of sewage sludge ash with vegetable oil at a low temperature, the team have developed bricks that have an ultra-low, even negative carbon footprint. The production process doesn't even produce waste or use water. An independent estimate suggests that if these bricks are produced on a large scale they have potential to reduce greenhouse gas emissions by up to 160 and 120 per cent when compared traditional concrete blocks and clay bricks. A number of samples have already passed crucial industry testing for freeze-thaw and compressive strength.

Encos has recently commenced the manufacture of products for testing and trials from its pilot plant on Yorkshire Water's site at Knostrop, Leeds. This most unusual collaboration could effectively mean that some of our own waste would go to use as a component of an ecologically friendly building material. The company's patented method for manufacturing carbon-negative masonry products from waste materials is the result of research carried out by Dr John Forth and his team at the University of Leeds' School of Civil Engineering.

PROPERTIES

Binding

Carbon-negative

Good fire, freeze-thaw and compressive strength

Solid

Sustainable

APPLICATIONS

Alternative to clay bricks and concrete blocks

Construction materials

Masonry

Waste management

INFO

encosltd.com

www.leeds.ac.uk

Leeds University spin-off company Encos Ltd and Yorkshire Water's unusual bricks. Image courtesy Encos and Leeds University.

CATALYTIC PLASTER

PROPERTIES

Breaks down dirt and odour

Converts pollutants to harmless particles

APPLICATIONS

Architectural surfaces

INFO

www.redstone-usa.com

Along with the regular functions of plaster, this material is also able to reduce pollutants in the air and is stain resistant. One such Catalytic Plaster product is redstone LUNO™, an odour-dispelling, non-staining titanium dioxide-based plaster wall covering. At a molecular level, it works by releasing negatively charged electrons from the titanium dioxide crystals upon exposure to light, which in turn allows oxygen radicals to form in the plaster. This oxygen is capable of breaking down organic molecules, which are the cause of unpleasant odours, moulds and stains on the plastered walls.

Additionally, the photo-catalytic reaction between the titanium-dioxide covering and natural UV light causes the plaster surface to become hydrophilic, creating a microscopically-thin film of the liquid which stops any dust or dirt and subsequent stains permanently adhering to the wall so they can be easily swept away.

SLIM BRICKS

PROPERTIES

30 per cent lighter than regular bricks

Can be installed like regular bricks

APPLICATIONS

Architecture

INFO

www.wienerberger.nl

An ability to make bricks lighter would save the construction industry money: in materials, in manufacture and in transportation. SlimBricks, a product manufactured by Wienerberger in the Netherlands, are simply that—slimmer bricks. Although the material make-up, the construction method and the manufacturing method remain the same as for regular bricks, the reduced weight has the potential to save both time and money.

SlimBricks are an example of where a redesign of a current and widely used object has the potential to revolutionise the industry in which it is used and how materials innovation is at the heart of this process.

The bricks are load bearing but thinner than regular bricks, reducing material cost in manufacture. Making the bricks thinner allows for a larger layer of insulation in the wall cavity, thus helping to make the built structure more energy efficient. Image courtesy Wienerberger.

TOP
Glass-filled Nylon's texture is similar to tough plastics, but, because of its Glass inclusions, it is much stronger, making it an ideal engineering plastic. Image courtesy PAR Group.

GLASS-FILLED NYLON 66

Nylon 66 GF 30 is a 30 per cent glass fibre reinforced semi-crystalline engineering thread with high strength and varied applications. It is widely used in the automotive industry, but is also used in the manufacture of all sorts of high functioning plastic products such as spark-plug parts, friction bearings, rollers, gears, piston guides and pulleys.

It is very strong and rigid and resistant to many oils, greases, diesel and cleaning fluids. It has good dimensional accuracy so can be made into precise shapes, and resists deformation when exposed to moderate heat and mechanical stresses. It is the kind of material we rarely notice, as it exists within the workings machines and objects, yet it can be found in nearly every modern car on the planet.

PROPERTIES

Good dimensional accuracy

Easily machined

Easily bonded

Temperature Range: from -40 to +110 degrees Celsius or -40 to +230 Fahrenheit

APPLICATIONS

Engineering parts

Piston guides

Bearings

Plug parts

INFO

www.par-group.co.uk

FIBREGLASS

Fibreglass is a much used composite material composed of glass fibres and resin. The glass fibres are very strong in tension and compression but have comparatively weak shear strength. The way the materials are aligned and combined in Fibreglass is therefore what lends the material its good strength and flexibility across all axes. The fibres are set in a variety of angles across each other and when soaked with the resin and hardener mixture, form a stronger material, reinforced in all directions.

Fibreglass is used in everything from architectural applications to wind farms. It is extremely resistant to weathering and erosion and is therefore the material of choice for speedboats and docks, which are continually exposed to seawater.

PROPERTIES

Strong

Tough

Resistant to corrosion

Light

APPLICATIONS

Boats

Architecture

Product design

Wind turbines

INFO

www.stuartpease.co.uk

CARBON FIBRE COMPOSITES

MATERIALS FOR RACING

In the 1980s, engineers in Formula 1 racing revolutionised the construction of their cars by using Carbon Fibre to replace metal in building body work. The material offered dramatic improvements in weight reduction and strength enhancement, as well as being easily shaped. Nowadays, Carbon Fibre is also a key component of the car's chassis. The innovative application of Carbon Fibre into these aerodynamic machines galvanised the world of motor racing. Great effort is invested in developing the Composite construction and mechanical properties of Carbon Fibre with many different types and treatments yielding a variety of material performances.

Taking their extensive knowledge and research in the automotive industry, Audi Concept Design have recently developed skis to produce Audi Carbon Skis. They are made from a wooden core coated with aluminium and titanium and fine layers of precisely laid Carbon Fibre. The use of Carbon Fibre makes the skis around 200 grams or 0.44 pounds lighter than comparable ones. The way that the individual Carbon Fibre strands are laid against each other, and the form of the ski, lends the ski stiffness while simultaneously minimising its uneven surfaces and ice.

OPPOSITE
A section through the front of a Formula 1 racing car. The Carbon Fibre bodywork makes up only a relatively thin section, such is its strength. The shell is then filled with foam to make the structure rigid while remaining incredibly light.

ABOVE
The front of a Formula 1 car. The Carbon Fibre Composite is sprayed, but cutting though it would reveal the same structure as shown left.

RIGHT
Audi concept skis use the kind of technology developed through Formula 1. The Carbon Fibre pattern is used as both structural component and for its aesthetics. Image © Audi.

FERROFLUIDS

PROPERTIES

Magnetic

Alterable

Extremely smooth surface

APPLICATIONS

Sculpture

Medicine—MRI Scanners

Lubrication in mechanical engineering

INFO

wood.phy.ulaval.ca

sachikokodama.com

Ferrofluids, or magnetic liquids (first invented in NASA laboratories in the 1960s) are composed of ferrous nanoparticles suspended in a liquid, typically water or organic solvents. When a magnetic field is applied, the viscosity of the liquid increases in response to the magnetic force and the lines of the magnetic field are rendered visible as the liquid moves and form patterns.

Today, Ferrofluids are used in a variety of scientific and engineering contexts. For example, some luxury cars have the material in their suspension system, enabling the driver to control the hardness of the ride by turning a knob that affects an electromagnet, in turn altering the viscosity of fluid. Ferrofluids are also used as lubricants due to their friction-reducing capability. In medicine, special types of ferrofluids can be used to increase contrast in MRI scanners by injecting them into the body, allowing for the detection of diseases such as cancer.

Artist Sachiko Kodama has also used Ferrofluids to create dynamic sculptures that rely on the material's fluidity and magnetic properties, only remaining three-dimensional for as long as they are magnetised: when the field is gone, the liquid loses its form and collapses.

Researchers at Laval University have pioneered an innovative use of Ferrofluids, using the morphing and reflective qualities of the material to create liquid, deformable mirrors. The surface of the liquid is extremely reflective and, using an electromagnet, the mass can be shaped to form perfectly smooth convex mirror forms. These mirrors have potential application in telescopes and other technical optical applications.

LEFT
Laval University's homemade Ferrofluid coated with a MeLLF. The image shows the liquid deformed by the presence of a magnetic field of a permanent magnet located under the container. Image courtesy Laval University.

OPPOSITE
Detail of Sachiko Kodama, *'Protrude, Flow 2008'* exhibited at the Museo Nacional Centro de Arte Reina Sofia (MNCARS) in Madrid in the Maquinas & Almas (Souls and Machines) Exhibition. Image © Sachiko Kodama, photograph, Mario Martin.

RHEONETIC MAGNETIC PASTE

PROPERTIES
Dark grey liquid
Magnetic

APPLICATIONS
Energy dissipating for shocks
Fluid dampers
Industrial suspension

INFO
www.lord.com

Rheonetic Magnetic Pastes and Liquids are materials capable of changing from solid to liquid and back again in milliseconds!

Made from micron-sized magnetically polarisable particles suspended in a carrier medium, the particles form chains when a magnetic force is applied. The strength of the magnetic field affects the number of chains formed—the stronger the magnet, the denser the chains formed, the thicker the material becomes.

Discovered in the 1940s, the technology is getting a new lease of life, replacing hydraulic systems in heavy vehicles, creating a smoother ride for the driver. LORD Corporation has revived it, coupling the paste with electronic micro controllers, creating a system for truck seating capable of controlling shock from some of the roughest trucking roads in the world. Using electromagnets with varying currents, their Motion Master Ride Management System is able to designate three settings that alter the ride at the touch of a button—a much-needed upgrade from the present hydraulic system which has a much slower response time.

SOFT MAGNETS

PROPERTIES
Soft
Magnetic
Shock absorbing

APPLICATIONS
Switches
Toys
Seals

INFO
www.taica.co.jp
www.inventables.com

The inclusion of magnetic particles in Taica's Alpha GEL render this product magnetic. It is soft to the touch and can be squished, making it the perfect material for silent drawers or cupboard closer mechanisms. The material's hardness, size and shape can be customised because, when up-scaled for industry, it will be injection moulded. The material is very new, which means that its potential uses are still being developed, from inclusion in magnetic toys for children to switches and magnetic seals.

Taica's Magnet GEL. The magnetic strength can be altered to suit the application, as can size shape and elasticity. Image courtesy Taica.

OPTIC FIBRE FABRIC

Luminex is a fabric with integrated Optic Fibres. The bunched ends of the Optic Fibres can be arranged to meet small LEDs, spreading light throughout the fabric. Because of the light-weight nature of the textile and the thin fibres, the light appears to shimmer across the surface.

The simple effect of transmitting light across a fabric can be integrated with sensors and microelectronics, creating smart garments able to transmit signals about the body that wears them if coupled with sensors.

PROPERTIES

Light-weight

Colourful

Light transmitting

APPLICATIONS

Garments

Decorative textiles

INFO

www.luminites.co.uk

Optic Fibres integrated in a blue fabric. Light is emitted from the cut ends. Image courtesy Luminites.

PAPER MADE FROM STONE

Terraskin is made from two parts calcium carbonate, the stone of its title, and one part Polyethylene binder though looks and feels just like regular paper. The process of production, however, wastes no trees, does not use water or bleach and produces papers that are more durable than those made from tree pulp.

The paper offers exceptional printing capabilities—because it is not made from fibres, ink doesn't get absorbed, which means that the paper can be decorated with very crisp images, perfect for specifically branded products.

PROPERTIES

Smooth

Easy to print

Tough

Recycled

Recyclable

APPLICATIONS

Packaging

Branding

Bags

INFO

www.terraskin.com

STIFF FABRIC ORIGAMI

PROPERTIES
Light-weight
Foldable

APPLICATIONS
Sculpture
Product design

INFO
www.foldtex.com

Using the principles of origami, but with fabric rather than paper, designer Timm Herok has created Foldtex. It is a light-weight foldable board composed of two layers—one flexible, the other stiff. Used as a sandwich material, a hinge can be created by making an incision into the stiff layer and exposing the flexible layer and allowing it to move.

Foldtex can be customized to create a range of textures and effects by adding individual surface layers to the flexible layer in different materials, all from simply connecting a flexible material to a stiffer material and applying a series of cuts. As a design project it is a clear example of the benefit of Composites—where one material is combined with another to create a new material with the combined properties of both, in this case stiffness and flexibility, capable of holding geometric three-dimensional patterns.

Details of patterns that can be achieved with Foldtex. The resulting material can be stretched or made into products with specific capabilities. More folds will render the material more flexible. Images courtesy Timm Herok.

THE IDEA OF A TREE

MICHER'TRAXLER

Designer duo mischer'traxler (Katharina Mischer and Thomas Traxler) have made a machine called "Recorder One" comprised of plexiglass and stainless steel, and some clever electronics, which is able to construct furniture from twine and glue.

Addressing the very relevant theme of environmental change, the machine reacts to the hours of sunlight in a day as well as subtle shifts in the environment as fleeting as shadows. The machine works only in the daytime, from sunrise to sunset. 'The idea of a tree' project enables the transience of nature to be recorded and preserved in an item of furniture, be it a bench, container or lamp.

Threads of viscose pass through a basin of solar sensitive paint that acts as a dye. Having also been pulled through glue, the now coloured natural fibres are then wrapped round a solar-powered plastic mould that rotates to give the piece its final form. An old-fashioned stamp is sealed to the finished piece to show the day and place of manufacture, archiving the original location. Paler colours and larger pieces are indicative of long days with stronger, more intense sunshine, while darker colours with less layering indicate short, darker days, thus representative of the corresponding seasons and weather. "The speed of this process is determined by the intensity of the sun's rays", they write, just as a tree's growth and its leaves photosynthesising are affected by the day's conditions.

LEFT
The machine, powered by a solar panel, which determines the speed at which the tube is turned, pulling the string through the dye and glue and thus recording the sunlight on the day the work was made.

OPPOSITE
Lightshades and Stool. Each piece is stamped with the time and place of its making. Despite being made of only string and glue, these thin-walled pieces are structurally sound, such is the strength of this material combination. Images courtesy micher'traxler.

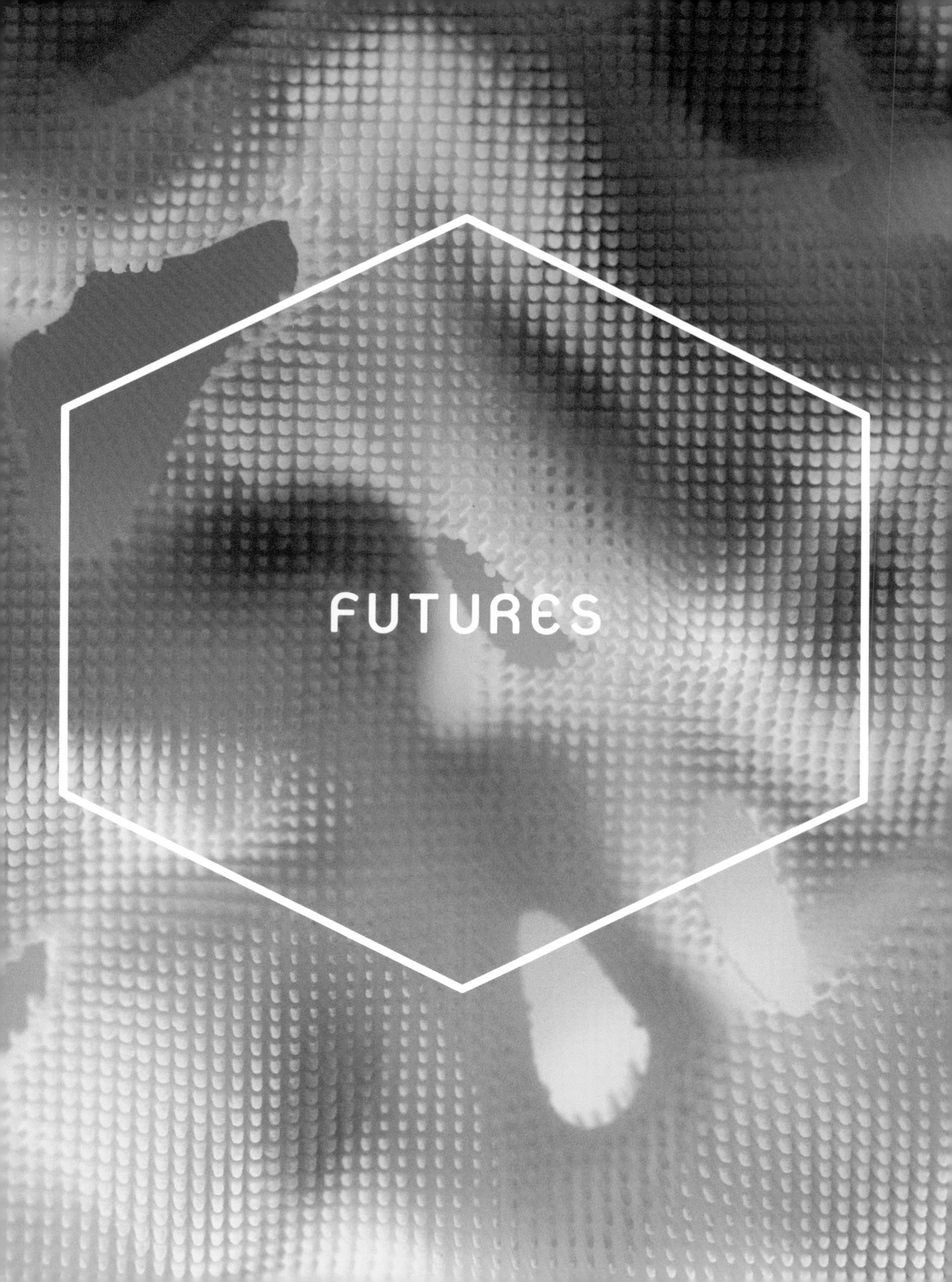
FUTURES

Many possible futures present themselves when one starts to imagine what might lie ahead in the world of materials. New processes, combinations, techniques and properties await discovery and exploration by practitioners from a wide range of disciplines, each interested in various aspects of materiality. Nanomaterials are already with us, three-dimensional printing is a reality and biological systems are currently being manipulated to produce new forms of matter. In other words the future is already here, but we are only at the beginning of understanding what we can do in it.

The concept of biomimicry has already emerged as a powerful way of tackling new challenges. Famous examples like Velcro, Self-cleaning Glass and the sharkskin-inspired swimming costumes worn by athletes in the 2008 Olympics, have all taken their inspiration from the natural processes. However, this is only the beginning. When mankind is faced with an engineering problem, it often emerges that this problem has already been faced by Nature, and conquered in subtle and unexpected ways. Our efforts at turning sunlight into energy are only moderately successful at best, yet every leaf on every plant the world over is performing such a function with efficiency and sustainability all of the time. Composite materials in the natural world exhibit a strength that seemingly defies logic, given their constituent parts. The concept of self-healing is also fundamental in Nature, but is currently lacking in our technologies, though a few prototype examples are emerging, with Self-healing Concrete being one of them. By looking at, understanding and taking inspiration from some of the processes and methods Nature uses to achieve remarkable material feats, some possible material futures can be revealed.

Biomaterials are made all the more incredible by the fact that they are naturally occurring: they self-assemble spontaneously from atomic and molecular building blocks, with exquisite detail at every length scale. This idea of taking control over the fundamental building blocks is central to nanotechnology, a scientific movement which is beginning to change the way we think of materials. Many traditional manufacturing techniques rely on taking bulk quantities of material, then cutting and shaping them into a final structure. Through nanotechnological methods, we can approach materials manufacture from the other direction: we can grow them from the bottom up, one atom or molecule at a time. This length scale is of course beyond the tactile grasp of a human being, and no machine exists which can reliably build nanostructures at anything like an industrially viable scale. Instead, we rely on chemical self-assembly to build nanostructures. So, like in the biomaterials mentioned above, we can take very specific chemical ingredients, in just the right chemical environment, and we can make matter self-assemble. A common example is nanoparticles, which are emerging as a very exciting technology in diagnostic and therapeutic healthcare applications. Examples include polymeric nanoparticles used to carry drugs to targeted areas in the body, Gold Nanoparticles for ever more accurate biosensing and imaging, and fluorescent quantum dots used to probe the mechanisms at work inside animal cells.

Bottom-up materials assembly may well be one of the directions in which three-dimensional printing develops. At present, sophisticated processes of three-dimensional printing enable objects to be created additively, with the form built up in a succession of thin extruded or sintered layers in a range of materials like gold, nylon or chocolate. As technologies enable materials to be deposited within the printing process in smaller and smaller amounts, the resolution and resultant complexity of the forms that can be printed increases. Contemplating the implications of such advances, it is possible to imagine a future in which not only can the form of and object be printed, but the structure of the material itself, thus the possibility of assembling new types of matter. Here, properties that are not the direct properties of the material extruded or sintered by the printer, could be generated through small scale structural design. For example, a ridged polymer could be laid down in a manner that produces a structure that gives the finished form elasticity, even though the polymer itself is not elastic.

It is always difficult to imagine the creation of things that have never existed before, especially when the tools of their creation have yet to be imagined, but similarly, the development of new tools and techniques for materials manipulation enables a whole new sphere for the imagination of materials. It is highly unlikely that humanity will discover new elements that will sit in a stable way on the Periodic Table, but we may very well develop new ways of manipulating the combination of those fundamental building blocks. The boundaries between the animate and inanimate could blur, and in such a way we could start to imagine things that have never existed before. Whatever the future holds, new and creative ways of looking at and using the materials we already have will be vital. How to be resourceful and efficient in the use of pre-existing materials and operate with a unified approach to the material world is a clear challenge for any future.

PH and ZL

CASE STUDY

DESIGNING HYBRID MATERIALS

BREAD LTD

BREAD Ltd is a collective that operates as a cross between a design consultancy and an incubator, both developing their own projects and working with clients. They came together because they felt that the way design operates commercially was pretty limited in harbouring research. Sarat Babu, the founding member of the collective writes: "Three years ago I asked myself the questions: 'what would happen if an object could have different properties within its volume? Could it be soft in one place and stiff in another? Could it be heavy and light? I started exploring ways of making these kinds of objects, by adapting material structure." With this, an ongoing project called "Designing Variable Materials" was born.

Building materials using Additive Layer Manufacturing enabled Babu to develop materials with different properties, such as more weight in one part versus greater insulating properties in another. Much like one might weight a dice to fix the outcome of the roll, an object with a specific function can have that function enhanced or adapted through altering not only the geometry, but the microstructure. For example a medical splint could be tailored to the wearer's own body, supporting and stretching where necessary, and a dice could be fixed to always land on the six.

This method of designing materials turns the design process on its head. Commonly designers devise a shape then pick a suitable material for

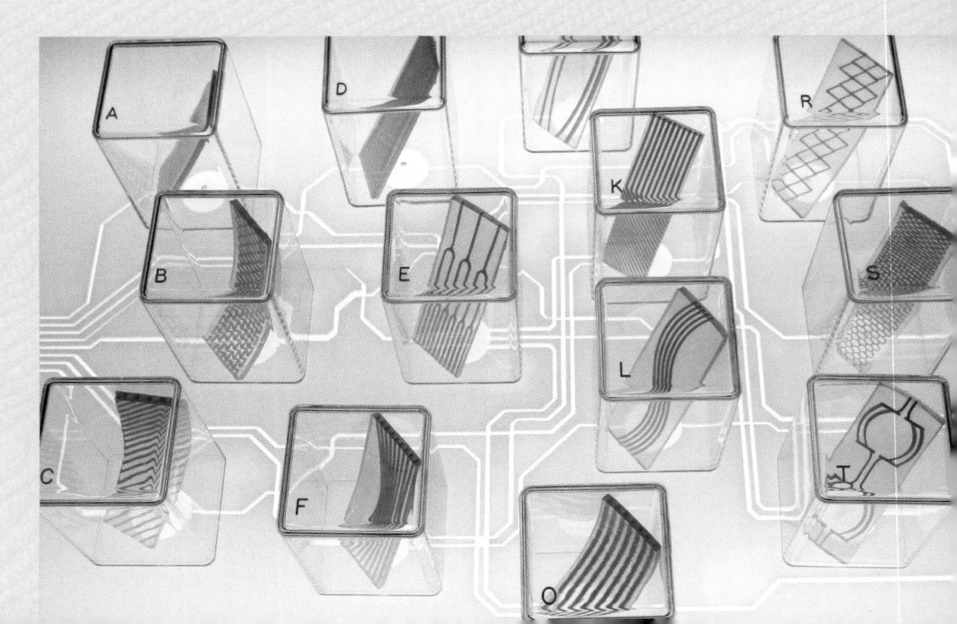

RIGHT
A series of sample materials with different functions. The project was supported by the EPSRC and the Royal Commission for the Exhibition of 1851. The technology is additionally in immediate commercial development internally and with various commercial partners.

OPPOSITE
An auxetic foam printed using additive layer manufacturing. Using this technique it can be engineered to compress more one side or the other, or if pulled, the structure can be engineered to expand. In this way, new materials can be developed for very specific tasks with highly specified internal structures.

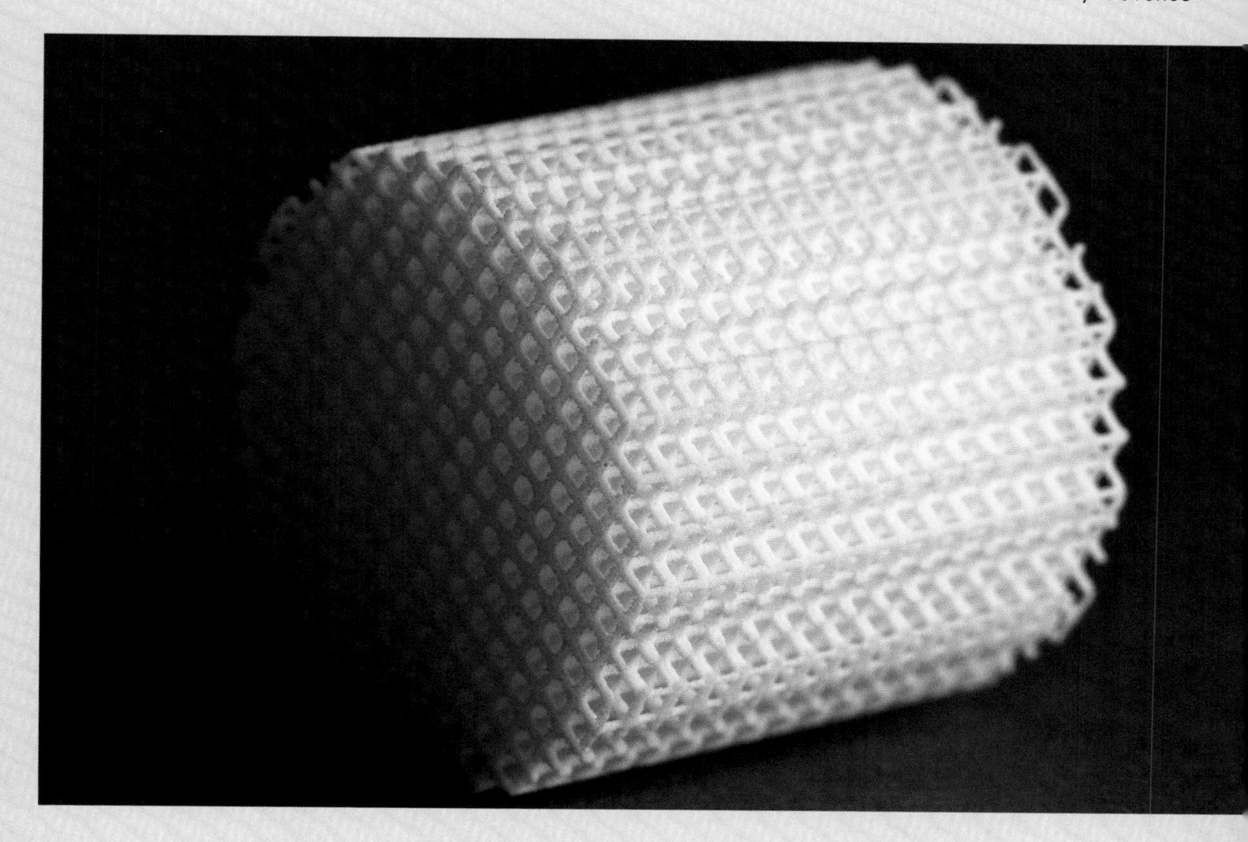

it or else decide upon a material and design a suitable form. The process is very linear. With variable materials, however, Babu wasn't choosing a material, he was designing it, so considering the object's specific function and behaviour becomes very important, not just in general for use with every user, but specifically for each different user—in essence, smart and extremely efficient objects.

Babu has been working on these structures on a BREAD Ltd sponsored doctoral program at University College London. He has been developing biomedical applications for a while now, building materials with specific functions, which enables him to closely mimic the behaviour of tissues, which could lead to vast improvements in synthetic replacements.

The potentials of the process are huge however. As the process gets cheaper even the most commonplace items could be made through Additive Layer Manufacturing. As well as creating smarter objects that could also change worldwide manufacturing as we currently know it, reducing waste, traveling distances and need for raw materials.

GOLD NANOPARTICLES

PROPERTIES

Biocompatible

Nanoengineered

Non-reactive/stable

Robust against chemical attack

Selectively latch onto protein markers—which means they can be used to detect breast cancer

Gold particles exhibit a large variety of colours

APPLICATIONS

Stained glass

Medical advances

Light scattering microscopy

Stabilising scaffold for drugs

Biosensors

INFO

www.chem.purdue.edu

news.uns.purdue.edu

www.stevensgroup.org

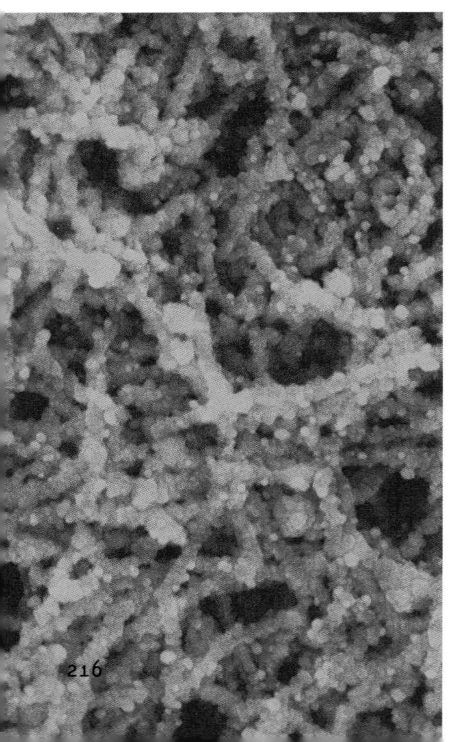

Although Gold Nanoparticles have been used in stained glass for centuries, they are currently making headlines for their role in various amazing medical advances. Although we are all very familiar with gold as a bulk material, when it reduced down to the nanoscale it exhibits very different behaviours, and it is these which can be harnessed in biomedical applications. Importantly, as the particles are nanosized, they can enter the body and interact on a molecular level, but they are robust against chemical attack so they remain stable and operative.

Solutions of Gold Nanoparticles exhibit a large variety of colours depending on the size and proximity of the particles suspended in the liquid. This change of colour can be exploited to make new medical tests which are required to detect the presence of a certain chemical or type of cell. For example, a group at Imperial College London is developing biosensors based upon Gold Nanoparticles that causes solutions containing these nanoparticles to change colour in the presence of particular enzymes.

At North Carolina State University scientists have rejuvenated a failed HIV drug, first tested in the 1990s and subsequently abandoned for its then bad side effects, by incorporating Gold Nanoparticles. The nanoparticles acted as non-reactive stabilising scaffold for the TAK-90 drug and significantly improved its performance.

In 2007, Purdue University also discovered a way to detect breast cancer using Gold Nanoparticles. Using specially shaped Gold Nanorods with specific antibodies bound to the surface, the gold particle assemblies selectively latched onto protein markers for breast cancer, which was then detected by light scattering microscopy techniques.

The use of Gold Nanoparticles in medicine has only just started to be fully explored; however their huge promise is evident. However, there are many careful testing procedures which must be fulfilled, and legislative hurdles to be jumped, before the use of nanoengineered materials becomes mainstream in biomedicine.

LEFT
Gold nanoparticles create visible-light catalysis in nanowires. Image, Argonne National Laboratory.

OPENING PAGE
A detail of Cilia by Sarat Babu and Richard Beckett for Future Thinking at the Surface Design Show, 2012.

GRAPHENE

Graphene has received a huge amount of press attention in the last couple of years. It was first identified and man-made by Professor Andre Geim and Dr Kostya Novostev, an act that subsequently won them the Nobel Prize in 2010 as Graphene has the potential to completely revolutionise technology as we know it.

Graphene is simply a single atomic layer of carbon graphite, the material which pencil leads are made from (a stack of three million sheets would be only one mm thick!). As you draw with a pencil, the friction between the graphite and the paper acts to pull off these layers, leaving them trailing behind and stuck to the paper. When one of those atom-thick layers is isolated as pure Graphene, it constitutes a material which displays some extraordinary properties.

Testing has shown Graphene to be one of the strongest materials ever measured, with a breaking strength 200 times greater than structural steel, and also as one of the most conductive, far more conductive than silicon.

It is also incredibly versatile. Likened in its vast potentials to something as ground-breaking as discovering plastics, its practical applications could be vast in number; the variations of uses, from inclusions into Composites to increase strength and conductivity, to use in sheet form as a digital screen, could potentially revolutionise technology. Jim Tow of Rice University has said with Graphene technology "you could theoretically roll up your iphone and stick it behind your ear like a pencil".

A huge amount of money is being poured into Graphene research by governments and the likes of IBM and Nokia, therefore the rate of development is very fast. It is already being slated for use in photovoltaic cells as it has the extraordinarily low resistance of a superconductor, but at room temperature rather, making it very energy efficient. It is also surprisingly elastic for a crystalline structure.

Most amazingly, it is a material that, at only an atom thick, can be seen with the naked eye and held in your hand.

PROPERTIES

Lowest resistance of any material at room temperature

Strongest material—100 times steel

Highest intrinsic mobility—100 times that of silicon

Highest current density—a million times that of copper for electric circuits

Best conductor of heat

Has the low resistance of a superconductor

APPLICATIONS

Composite material for strength and conductivity

Digital screens

Photovoltaic cells

INFO

www.graphenetechnologies.com

www.tmworld.com

www.graphene-info.com

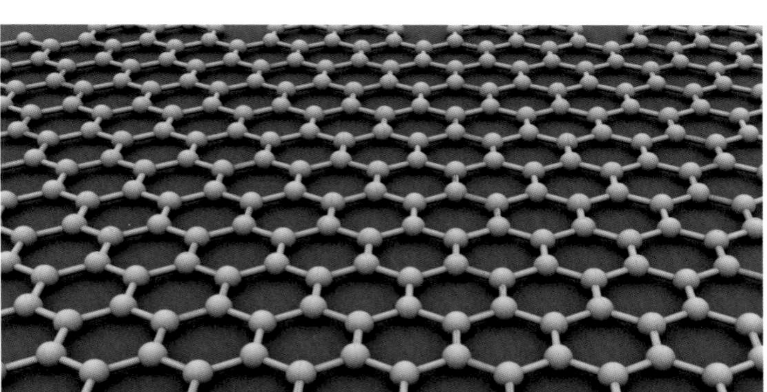

A graphic rendering illustrating graphene's hexagonal structure andstrong bonds between the carbon atoms. This two-dimensional structural arrangement gives graphene its characteristic 'chicken wire' appearance.

CASE STUDY

CATALYTIC CLOTHING

NANO-SCALE POLLUTANT REMOVAL

Catalytic Clothing is a collaborative project between two specialists in their respective fields: Helen Storey MBE, an artist, designer and professor of fashion and textiles, and Tony Ryan OBE, a chemistry professor and nanomaterials specialist. Between them they have created a harmless additive that can be applied to existing clothing during a normal laundry cycle which reduces air borne pollutants.

The process uses a nano-engineered photocatalyst to break down the pollutants in the air that come into contact with the treated cloth. Light energy absorbed from the sun frees electrons in the material, which then react with molecules of water in the air, breaking them down into two radicals that are hugely reactive with air pollutants. In the reaction, the radicals bond with the pollutants, breaking them down into harmless chemicals that can simply be laundered away and filtered out in our ordinary water filtration systems.

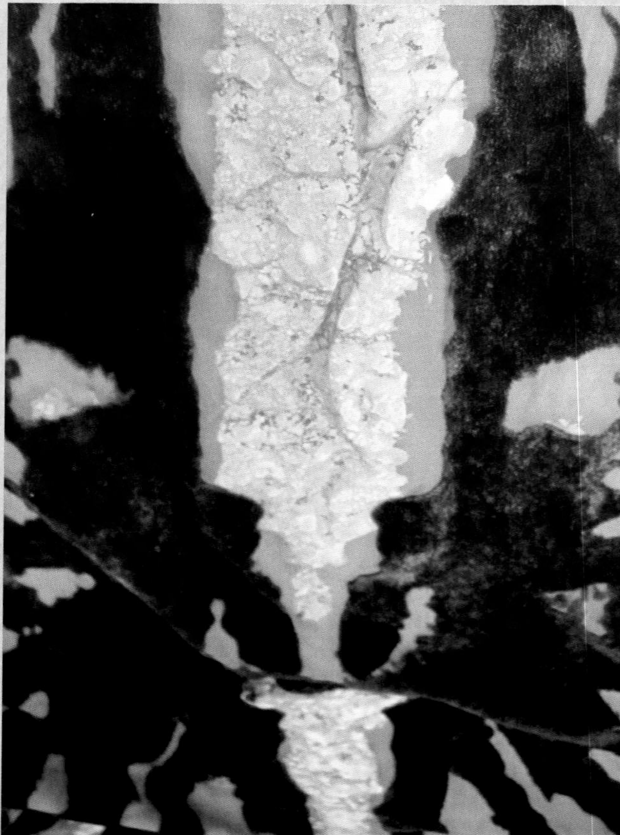

OPPOSITE LEFT
Erin O'Conor as Herself, still from Catalytic Clothing film. Image © Adam Mufti.

OPPOSITE RIGHT
Detail of Herself textile. Image © Trish Belford.

ABOVE
Field of Jeans, Newcastle, 2011. Image © Gavin Duthie.

RIGHT
Field of Jeans Signage, Newcastle, 2022. Image © Chris Auld. Images courtesy Catalytic Clothing and Helen Storey.

Using a catalytic converter to eradicate pollutants is not a new technology—some existing paints, cements and other building materials already reduce air borne pollutants—but what makes Catalytic Clothing unique is the ease of application and its potential for mass use—on all clothing, both existing and yet to be manufactured.

The photocatalyst can be delivered to fabric within a standard product such as a fabric conditioner. The active part is packaged within a shell that attaches to the surface of clothing during the wash cycle. In a few cases, the pollutants are attached without breaking down, but tests have shown that they can be washed off and will be removed safely through the water purification system.

PALLADIUM GLASS

PROPERTIES

Outstanding strength and toughness, simultaneously

A microalloy made from nobe metal with a small fraction of silver

Amorphous structure

Optically and electrically akin to metal

APPLICATIONS

Industrial applications where extremely high damage tolerance is required

INFO

www.lbl.gov

media.caltech.edu

This new type of glass has the potential to be one of the best materials on earth for exhibiting both strength and toughness. Although there are many materials that are tougher (more energy is required to fracture them), and many materials are stronger (they can take more force before deforming), Palladium Glass is special because it is simultaneously both strong and tough, which is an unusual combination.

Palladium Glass is a microalloy made from palladium and silver metals. However, instead of exhibiting the polycrystalline structure typical of metals, it has an amorphous structure which is characteristic of typical glasses. In forming this structure with these metals, a material is obtained which harnesses the qualities of both types of material and overcomes the inherent brittleness of glass. It does not look like a typical glass and only takes its name because of the amorphous structure. Optically and electrically, it behaves like a metal.

Marios Demetriou at the California Institute of Technology states that the “study demonstrates for the first time that this class of materials, the metallic glasses, has the capacity to become the toughest and strongest ever known”. This is particularly exciting because it pushes past the level of damage tolerance currently viable for a structural metal.

“When defects in the amorphous structure become active under stress, they coalesce into slim bands, called shear bands, that rapidly extend and propagate through the material,” says Demetriou. “And when these shear bands evolve into cracks, the material shatters.”

Metallic glasses were originally developed in the 1960s at the California Institute of Technology. It was thought then that they could never be tougher than the toughest steel. However, notches intentionally created in the material allow cracks to form without deforming the material. The distance each crack can travel within the material is blocked by the multitude of tiny shear bands that form, which is what strengthens the material.

LIGHT EMITTING NEAR INFRARED MATERIAL

Although invisible to the naked eye, near-infrared light makes up nearly 50 per cent of the light energy that reaches earth from the sun. Night vision devices work by converting near-infrared light into visible light, so that we can see it. At the University of Georgia, scientists have recently developed a new material that emits a long lasting near-infrared glow after only a minute of exposure to sunlight.

While this phosphorescent glow simply looks like any other glow in the dark product, it in fact has incredible potential to revolutionise medical diagnostics, create an undercover source of illumination not otherwise detectable at night and may have potential to create more efficient solar cells.

One of the most amazing applications is in the detection of cancer. It can be formed into nanoparticles that bind to cancer cells allowing doctors to visualise tiny, otherwise undetectable tumours, as infrared light is able to penetrate human tissue. Alternatively, when combined with a ceramic, or used as a paint pigment, objects can be made that are only visible to people wearing night vision equipment.

The material glows for around 360 hours, and can be activated by indoor fluorescent light. Extensive testing in seawater, in all weather conditions and even in corrosive bleach has not affected its ability to store and transmit near-infrared light. Zhengwei Pan's team at the University of Georgia are now working on using it to make solar cells more efficient by harnessing infrared light from the sun which is not usually captured by photovoltaic devices.

PROPERTIES

Stores and transmits near-infrared light

Activated by indoor fluorescent light

Harnesses infrared light from the sun

Glows for around 360 hours

APPLICATIONS

Very efficient solar cells

Medical diagnostics

Undercover source of illumination

INFO

www.uga.edu

chronicles.franklin.uga.edu

Researchers at the University of Georgia have developed a new material that emits a long-lasting near-infrared glow after a single minute of exposure to sunlight. By mixing it with paint, they were able to draw an image of the university's logo whose luminescence can only be seen with a night vision device. Credit: Zhengwei Pan/UGA

DRESSING MEAT FOR TOMORROW

IN-VITRO MEAT

In-vitro meat, also known as cultured meat, is the specific focus of a project by James King, a speculative designer, which investigates the potentials of recent advances in tissue engineering. In an era where meat is farmed in pens, King suggests that science has the potential to put the brakes on over-farming and even do away with the need for traditional livestock farming altogether. His work explores the possibility of manufacturing meat from a small sample of animal tissue—growing meat as a new material.

King, author of Dressing the Meat of Tomorrow, describes himself as "a speculative designer working to explore the implications of future biotechnology". He aims to encourage appreciation for the often-overlooked design element in biological science. Taking recent biological advances and working with them as a designer might, he speculates on what is possible and therefore devises scenarios that might become the norm in the future.

Due to the speculative nature of the work, the first example of King's in-vitro meat was not for a scientific experiment but for an art exhibition, part of 2003's "The Tissue Culture and Art Project", raising questions about its readiness for the consumer market. King's project is dismissive of what he describes as "the same old boring parts we eat today", leaving the imagination to explore what is now anatomically available and acceptable. No longer restrained by the actual interior geography of the animal in question, meat could be grown to an infinite size, in specific shapes or colours. In these experiments King questions not only what we will be eating in the future, but also what it will look like.

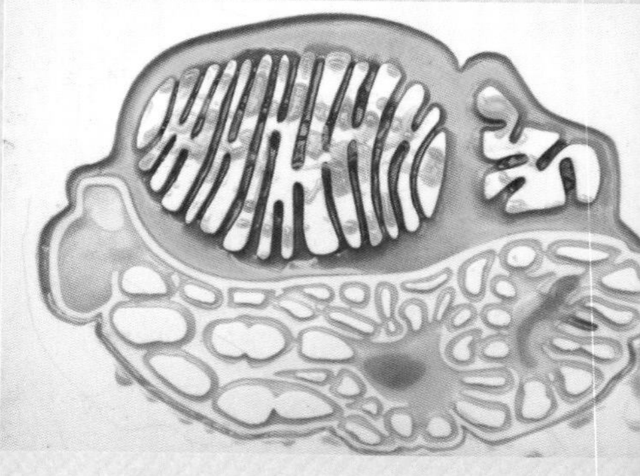

LEFT
Imagining what our plates of food might look like in the future.

OPPOSITE RIGHT AND ABOVE
The MRI Steak, described by James King as an anatomically complicated piece of meat. This project examines how we might choose to give shape, texture and flavour to this new sort of food in order to better remind us where it came from. Images courtesy James King.

PIEZO RUBBER

PROPERTIES

Flexible

Biocompatible

Produces a charge when flexed

Harvests the energy from a wearer

Removes the need for finite-life batteries

APPLICATIONS

Charge for phone or mp3 player

Pacemakers/other devices implanted in the body

INFO

www.princeton.edu

Piezoelectric materials generate a charge when struck or flexed. Researchers at Princeton University have developed a Piezoelectric Rubber that could be used to harvest energy from a wearer, producing power for electrical devices.

In a study reported in the journal *Nano Letters*, Michael McAlpine and his group reported their development of a flexible, biocompatible rubber film they called Piezo Rubber. It is composed on nano-sized ribbon-shaped piezoelectric crystals laid out on a flexible rubber substrate. When the rubber is flexed, the nanoribbons are mechanically stressed and produce a charge. This charge can then be harnessed to create an electrical current. As a wearable device, piezo-rubber may be used to charge your phone or mp3 player. However, there are suggestions that such a system could be implanted in the body to power devices such as pacemakers, thus removing the need to replace finite-life batteries.

LAB-ON-A-STICK

PROPERTIES

Instant diagnosis

Compact

Inexpensive

APPLICATIONS

Diagnosing disease

INFO

bioengineering.stanford.edu

A research team from UC Berkeley, Dublin City University and Universidad de Valpairaiso have made a revolutionary breakthrough in the field of microfluidics. The self-powered integrated microfluidic blood analysis system, or SIMBAS for short, is a Blood Lab-On-A-Stick—as simple to use as a litmus test and as effective as a full blood lab. It is a biochip independent of external inputs. By channelling almost all of the blood cells from plasma it can diagnose blood diseases such as HIV, malaria and tuberculosis in minutes.

The chip works by creating a pressure difference that pulls blood through the ultra small porous polymeric materials, separating the blood from the plasma. This occurs because the device is sealed in a vacuum, and once the seal is broken atmospheric conditions are restored and air molecules are pulled back into the material, creating suction.

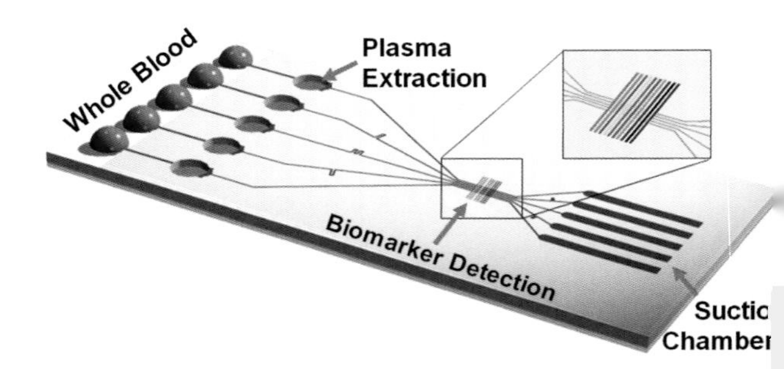

An illustration of the Lab-On-A-Stick bio-chip. The through flow, known as 'degas-driven flow' means the chip is as simple to use as a litmus test and as effective as a full blood lab. Image courtesy UC Berkley.

MANUFACTURED SILK

Shrilk is a biomimetic material developed by researchers at the Wyss Institute for Biologically Inspired Engineering at Harvard University, made using silk proteins and a long-chain Polymer called chitin. The material is designed to closely imitate the structure of arthropod chitin that is composed of layered protein and forms the tough outer layer of insect exoskeleton as well as shrimp, clam and snail shells.

For many, current plastics technology has reached an impasse, since it is heavily reliant on non-sustainable resources like fossil fuels. Scientists are therefore exploring the possibility of using Shrilk as an alternative in applications where petro-chemical plastics are ubiquitous. This represents the move to more eco-friendly and biodegradable alternatives to current plastic packaging, and its properties can be easily altered when used in a combination with other materials in the laboratory.

Shrilk is biocompatible, which means that the body does not reject it and cells can grow happily upon it. As a result, it is a material that provides potential for advancing the field of medical research, particularly in the area of regenerative tissue therapy. It is expected that Shrilk will be used to develop wound sutures that are capable of subsequent dissolution and actively stimulating the repair of the wound. It may also be used to produce coverings for wounds and burns, and to construct scaffolds that enable cells to grow into useful tissue and possibly even organs.

PROPERTIES

Clear/transparent

Biodegradable

Biosourced, yet low-cost

Tough (the dry material demonstrates a strength and toughness comparable to that of aluminium alloys)

When water is added, strength is deplenished, but the elasticity improves

APPLICATIONS

Replacement for plastics

INFO

wyss.harvard.edu

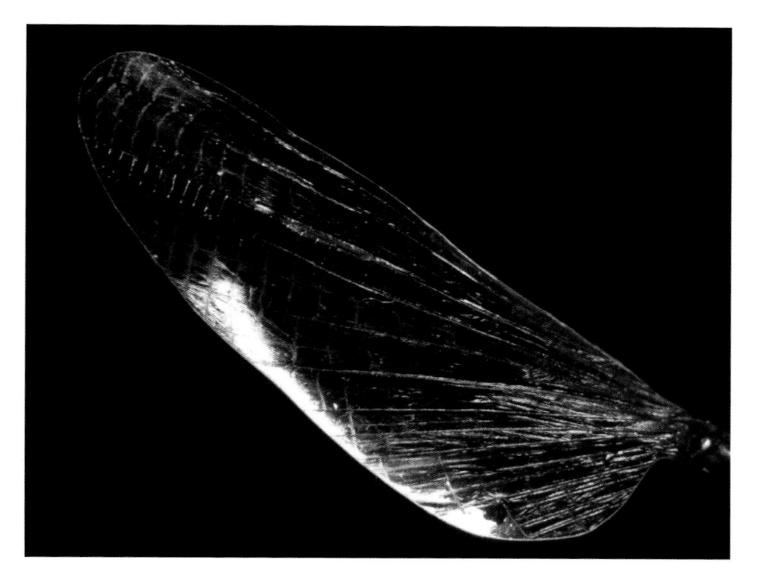

An insect wing provides the inspiration for the Wyss Institute's biomimicry. Image courtesy Harvard, Wyss Institute.

CASE STUDY

BODIES AS ENERGY CELLS

AFTERLIFE

Afterlife is a project by speculative designers James Auger, a research fellow at the RCA, and Jimmy Loizeau. The Afterlife process harnesses the chemical potential of the deceased body, converting it into usable electrical energy via a microbial fuel cell—a device that uses an electrochemical reaction to generate electricity from organic matter. The team are currently collaborating with scientists in order to make their speculative design possible. They wish to create electricity contained within a normal dry cell battery that can be used as a power source for a huge range of everyday electronic products. The project has the potential to turn a body into fuel, highlighting the material nature of our bodies and linking us to the finite fuels we burn freely for the power we use.

The project existed in two phases—the development and its applications. The second phase involved asking various people how they would make use of an Afterlife device. One man wanted to help with the little things in life, by having his life energy converted into useful objects, like becoming the battery for the garage-door opener or a hearing aid. In another case, a man wanted to achieve after his death what he never had in life—to fly a Spitfire MK1. He feels that this aeroplane connects with the soul so powering a plane is ideal and the phrase “remote controlled” takes on a whole new meaning. If used to power bicycle reflector lights, it could save lives too.

Many developing technologies inspire hope for the future, whether on global scale or in the improvement of day-to-day life. The Afterlife project shows how designers are considering the small-scale human stories and desires and incorporating them into technological advances that could actually impact on the large-scale consequences of consumption.

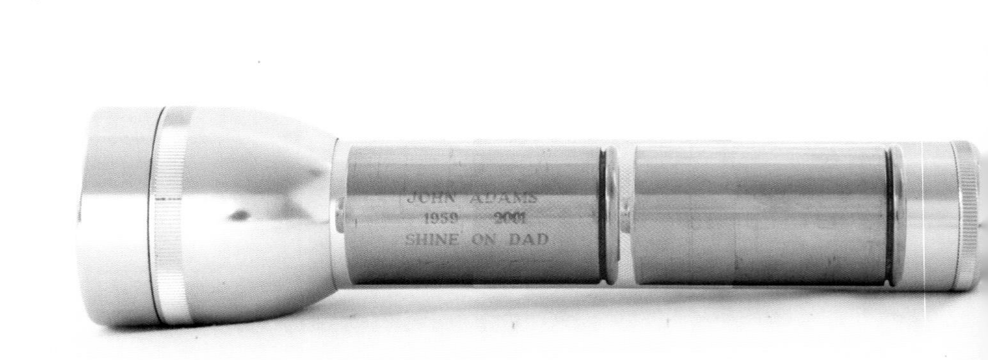

ABOVE
In phase II of the Afterlife project, James Auger and Jimmy Loizeau asked artists to develop ideas for the potential fuel created from bodies after death. Matt Karan's interpretation is for a person who wishes to be remembered more often than one use of a battery would allow for. He writes,"The deceased requests that a series of cells are manufactured, each with a random volume of electrolyte, so that the user of the cell never knows how long it will last. One may last a month, another a year. The deceased then, in death, continues to get the attention they so desired in life."

RIGHT
Afterlife Coffin rendering, depicting the microbial fuel cell and capacitor bank. Images courtesy Auger Loizeau.

GLOSSARY

A

Abrasion resistance
The extent to which a material resists erosion when in contact with mechanical actions.

Acid-free
Having a neutral or slightly alkaline pH.

Acid-resistant
Able to withstand degenerative damage by chemical reactivity or solvents of low pH values.

Alloy
A homogenous mixture of two or more metals.

Anisotropic
The properties of a material are not identical in all directions.

Annealing
Glass or metal heated and slowly cooled in order to harden and decrease brittleness.

Anodising
An anodic coating on the surface of particular metals, thickening their oxide layer, that allows pigmentation and electrical insulation.

Aramid Fibres
A class of heat-resistant and physically strong synthetic fibres.

B

Billet
A semi-finished solid metal that is double the thickness of its width, rectangular, circle or square.

Binder
Holds the particles within a material together and can provide powders properties like flow and malleability.

Biocompatibility
Works with or can be incorporated inside organisms without inducing harmful or poisonous effects on their immune system.

Biodegradable
Can be broken down over time by biological agents.

Biomimicry
The copying of natural processes or structures in man-made materials.

Brinell hardness test
Ascertains how hard a material is by placing it under a certain weight, measuring the results in terms of the Brinell hardness number (HB).

C

Calcining
To heat a material to a high temperature but below the melting or fusing point, causing loss of moisture, reduction or oxidation, and the decomposition of carbonates and other compounds

Calendaring
A finishing process for plastic that produces a smooth polished effect through the application of rollers at high pressures and temperatures.

Candela (cd)
Unit of energy emitted by light in one particular direction: luminous intensity.

Carcinogen
Cancer-causing agent.

Casting
Shaping of a material's final form by filling a mould with the molten material, where it is aloud to set.

Cellular ceramics
Contain a high level of porosity; the materials range from foams to honeycombs.

Cermet (ceramic + metal)
Materials made up of a mixture of ceramic and metallic components, combining physical properties of each.

Chemical milling
Removes material from surfaces using chemicals, also known as etching.

Coefficient
A quantitative expression of a physical property of matter, a constant factor in testing.

Compressive strength
A material is compressed until it reaches its utmost capability to withhold the force and is crushed.

Copolymer
Two or more different monomers reacting to form a polymer.

Creep
Over time, a solid material's deformation in shape or size due to external loads.

Crystal
A solid comprised of atoms, ions or molecules, ordered in a repeating three-dimensional pattern.

D

Ductility
The material's ability to change shape and not break.

E

E-Glass
Also named electric grade glass. It is a fibre with good electrical resistance, employed in laminating and reinforcing plastics. Commonly used as the reinforcing phase in fibreglass.

Elastic modulus
Can be referred to as either the modulus of elasticity, E or Young's modulus. Measures the material's degree of stiffness.

Elastomer
Rubber polymer that, as it is elastic, can bounce back.

Electrical conductivity
Measures the extent to which a material is able to conduct an electric current.

Electromagnetic interference (EMI)
Factors that interfere with the workings of electrical equipment, like radiation, other electronic devices or electromagnetic energy sources.

Emissivity
The efficiency of a surface in absorbing thermal radiation or emitting energy, compared to the emissivity of a theoretically perfect absorber, known as a black body.

Extrusion
The act of processing or shaping a material by forcing through a shaped hole or die.

F

Flexural modulus
Also known as Bending modulus, as it is an indication of a material's propensity to bend, measured by the proportionate levels of stress to strain.

G

Galvanised metal
The surface of iron or steel is layered with zinc in the aim of protecting against corrosive environments.

Gel
An internally cross-linked liquid that does not flow and therefore exhibits solid-like properties

Glass-ceramic
Crystallised materials that share properties of both ceramic and glass.

Glass
A non-crystalline solid that is three quarters sand, 20 per cent soda ash and five per cent lime, frequently mixed with metal oxides.

H

Hertz (Hz)
A unit designating frequency; measures the number of cycles per second.

Hydrophilic
Water loving material that easily absorbs moisture or dissolves in water.

Hydrophobic
Water hating, repels water.

Hygroscopic
Absorbs and retains moisture from surrounding environments. Common behaviour of cotton and food substances, like sugar and salt.

I

Impact energy
How much energy it takes to shatter a material, measured by various testing processes: the Charpy, Izod and Tensile impact tests as well as falling weight tests.

Infrared (IR) radiation-absorbing
The absorption of energy as waves in the electromagnetic spectrum, where wavelengths are longer than visible light but shorter than microwaves.

Ionomer
A copolymer linked by ionic bonds, sometimes known as a thermoplastic elastomer.

Ionic bonds
A type of chemical bond that involves electron exchange and subsequent binding due to electrostatic attraction.

Isotropic
Physical properties, such as conductivity or magnetism, are the same in all directions through the materials.

J

Joule (J)
Unit expressing electrical, mechanical and thermal energy or work. A Watt is equivalent to one Joule per second.

L

Light-emitting diode (LED)
A semiconductor that emits light, indicating the conversion electrical energy to light, used in lamps and digital displays.

Lignin
An organic and renewable raw material found in plants; an amorphous polymer that makes up to 30 per cent of wood and binds parts of straw.

Lumen (lm)
An amount of luminous intensity from a lighting source generating light in every direction equally, that is one candela in intensity per second.

Lux (lx)
Unit of illumination that is equivalent to one lumen per square centimetre.

Lyotropic
A material is called 'lyotropic' if it forms liquid crystal phases because of the addition of a solvent.

M

Microfibre
Very thin fibres, as fine as one decitex per filament.

Microspheres
Sometimes referred to as microparticles. Spherical shells made of plastic polymers or proteins with small diameters in the micrometre range.

Milling (machining)
The use of a tool that cuts away material.

Milling (powder technology)
Mechanical technique to change the shape or size of the material's particles or to cover one bit of a mixture with another, or to achieve an equal spread of the components.

Modulus
Mathematical description indicating the extent of a material's defining behavioural property.

Mohs scale
A scale invented by Friedrich Mohs for classifying 10 minerals based on relative hardness, determined by the ability of harder minerals to scratch the softer one listed below.

Moisture vapour transition rate (MVTR)
The measure of the time it takes water vapour to pass through a material.

Monoclinic (in context of zirconia/ monoclinic crystal system)
Of or relating to three unequal crystal axes, two of which intersect obliquely and are perpendicular to the third.

N

Newton (N)
Unit for force.

O

Opacifier
A chemical agent added to a material to make it opaque.

P

Pascal (Pa)
Unit of pressure.

Phase
The distinctive state of a material or the transition of matter; solid, liquid or gas.

Phosphorescent
A material property which allows it to emit light slowly after excitation ('glow-in-the-dark' materials), in contrast to fluorescence where light energy is emitted rapidly.

Piezoelectric material
Produces electrical energy from mechanical motion, or vice versa. The piezoelectric effect is the link between electrostatics and mechanics, which only occurs in non-conductive materials.

Plasticizer
A polymer additive that improves the elasticity and strength of the finished material.

Poisson's ratio
The ratio of lateral decrease to longitudinal increase of a sample of strained material that has been lengthened in the longitudinal direction.

Pyrolysis
Decomposition or transformation of a compound caused by heat.

R

Refractory ceramic
A ceramic material that is not damaged or deformed by extreme heat, used to make furnace linings.

Renewable resource
A material's supply meets demand due to biological cycles or sustainable processes, thus it does not run out.

S

Shear strength
Is the ultimate strength of a material subjected to shear loading. It can be determined in a torsion test where it is equal to torsional strength.

Shore hardness
Measures the give of elastomers by how susceptible they are to a Sceleroscope. The lower the number, the softer the material and the weaker the resistance.

Sintering
The bonding of adjacent powder particles in a mass of compacted powder, through heating.

Smart Materials
Materials with properties that can be significantly changed by external stimuli, many of which are designed to do so. Properties could include colour or shape and stimuli might be temperature, pH, electric or magnetic fields.

Stiffness
Refers to how much force a material can take before it cracks.

Strength
Refers to how much force a material can take before it deforms.

Stress
Force per unit area that influences the inner resistance of a material as well as deforming or straining the material itself.

Sublimation
The journey from a solid to a gas, bypassing the liquid stage.

Surface tension
Property of a liquid; the intermolecular cohesive forces at the surface allow it to resist an external pressure.

T

Temper
Temperature sensitive treatment that increases the strength and toughness of glass or metal materials through heating.

Tensile strength
The resistance of a material to a force stretching it in order to rupture it, measured as the maximum tension the material can withstand without ripping.

Thermal resistance
Measured by the difference between the temperatures of the material and the heat exerted upon it, therefore the material's ability to stop the flow of heat.

Thermal shock resistance
Susceptibility of ceramic and glass to split because of a fast drop in temperature.

Torsional stiffness
Ratio of applied torsion to the angle of twist.

Toughness
Explains the energy required to fracture or break something.

V

Vibration damping
Reduces the amplitude of vibrational noise.

Vickers hardness test
A universal measure of hardness; using a diamond indenter and changing loads for testing, a scale is available for many common solid materials.

Viscosity
Thickness of a liquid so that it is resistant to flow.

Vulcanisation
A chemical reaction causing a physical change in rubber, increasing elasticity and hardness.

W

Warp
In weaving, the threads that are held torsion on a loom, over and under which the weft yarns are woven.

Wear rate
In the sliding wear test, the total material worn away per unit.

Wear resistance
The resistance of a material to erosion due to the action of another surface, defined as the reciprocal the material's wear rate.

Weft
In weaving, the yarn that runs at right angles to those on the loom.

Y

Yield strength
Stress at which a predetermined quantity of the permanent alteration of a material begins.

RESOURCES

REFERENCE BOOKS

Ashby, Michael, and Johnson, Kara, *Materials and Design:* The Art and Science of Material Selection in Product Design, Oxford; Boston: Butterworth-Heinemann, 2009

Atena, Rossana, Pérez Arroyo, Salvador, and Kebel, Igor, *Emerging Technologies and Housing* Prototypes, London: Black Dog Publishing, 2007

Bell, Victoria Ballard and Rand, Patrick, *Materials for Architectural Design, London*: Laurence King Publishing, 2006

Beylerian, George M, and Dent, Andrew, and Moryadas, Anita (ed), *Material ConneXion: The Global Resource of New and Innovative Materials for Architects, Artists and Designers*, London: Thames & Hudson, 2005

Beylerian, George M, and Obsorne, Jeffery J, *Mondo Materialis: Materials and Ideas for the Future*, New York: The Overlook Press, 2001

Beylerian, George M, and Dent, Andrew, *Ultra Materials: How Materials Innovation is Changing the World*, London: Thames & Hudson, 2005

Brownell, Blaine, *Transmaterial: A Catalog of Materials that Redefine our Physical Environment*, New York: Princeton Architectural Press, 2006

Brownell, Blaine, *Transmaterial 2: A Catalog of Materials that Redefine our Physical Environment*, New York: Princeton Architectural Press, 2008

Brownell, Blaine, *Transmaterial 3: A Catalog of Materials that Redefine our Physical Environment*, New York: Princeton Architectural Press, 2010

Callister, William D, *Materials Science and Engineering: An Introduction,* New Jersey: John Wiley & Sons, 2006 (7th edition)

Campagno, Andrea (ed), *Intelligent Glass Facades: Material, Practice, Design*, Basel: Birkhäuser, 1999
Fernandez, John E, *Material Architecture: Emergent Materials for Innovative Buildings and Ecological Construction*, Oxford: Architectural Press, 2005

Gordon, JE, *The New Science of Strong Materials: Or Why You Don't Fall Through the Floor*, London: Penguin Science, 1991

Hudson, Jennifer, *Process: 50 Product Designs from Concept to Manufacture*, London: Laurence King Publishing, 2008

Kaltenbach, Frank, *Translucent Materials: Glass, Plastics, Metals, Basel*: Birkhäuser, 2004

Lees-Maffei (Author, ed), and Houze, Rebecca (Author, ed), *The Design History Reader*, Oxford: Berg Publishing, 2010

Lefteri, Chris, *Materials for Inspirational Design*, United Kingdom: RotoVision SA, 2006

Lefteri, Chris, *The Plastics Handbook*, United Kingdom: RotoVision SA, 2008

Norman, Donald A, *The Design of Future Things*, New York: Basic Books, 2009

Morris, Richard, *The Fundamentals of Product Design*, Lausanne: AVA Publishing, 2009

Peters, Sacha, *Sustainable and Multi-Purpose Materials for Design and Architecture*, Basel: Birkhäuser, 2011

Proctor, Rebecca, *1000 New Eco Designs and Where to Find Them*, London: Laurence King Publishing, 2009

Quinn, Bradley, *Textile Futures: Fashion, Design and Technology*, Oxford, New York: Berg, 2010

Seymour, Sabine, *Fashionable Technology: The Intersection of Design, Fashion, Science and Technology*, Berlin: Springer Publishing, 2008

Van Uffelen, Chris, *Pure Plastic: New Materials for Today's Architecture*, Berlin: Braun, 2008

Wessel, James K, *The Handbook of Advanced Materials: Enabling New Designs*, New Jersey: Wiley-Blackwell, 2004

REFERENCE WEBSITES

Accelerating Future
Blog written by Michael Anissimov, a science/technology writer and 'futurist' whose work focuses on emerging technologies such as nanotechnology and biotechnology and the central significance of technology for humanity's future.
acceleratingfuture.com

Physorg.com
Science, Research and Technology news forum—strong on smart materials and structures (piezoelectrics, magnetostrictors and shape-memory alloys).
physorg.com

Challenge of Materials Gallery, London Science Museum
Gallery specialising in the history and technology of materials. Encompasses interactive displays and installations.
sciencemuseum.org

Empa
Swiss Federal Laboratories—Materials Science and Technology for a sustainable future.
empa.ch

Institute of Making
Multidisciplinary Research Club for those interested in the made world: takes as its mission to provide all makers with a creative home in which to innovate, contemplate and understand all aspects of materials.
www.instituteofmaking.org.uk

Microscopy-UK
Monthly online magazine for enthusiast microscopists, with information resources, extensive image archives and educational links.
microscopy-uk.org.uk

Philips Design
Philips run the Design Probes research initiative—seeks to understand future socio-cultural and technological shifts, with a view to developing solutions for nearer-term scenarios.
design.philips.com/probes

The Institute of Materials, Minerals and Mining
Global network for professionals in the materials cycle—with links to networking, journals and training.
iom3.org

Transmaterial online
A companion to the Transmaterial series of books written by Blaine Brownell—concise resource aimed at architects and designers seeking to keep abreast of technological innovations.
transmaterial.net

Inventables
Online store and resource facilitating the sourcing of materials by designers, artists and inventors.
inventables.com

Mindsets online
Not-for-profit Design and Technology teaching resource. Reinvests surplus funds in education.
mindsetsonline.co.uk

Knowledge Transfer Network
Online networking site for industry innovators, researchers and academics—with a Smart Materials beacon.
ktn.innovateuk.org

Material Lab
Architecture materials lab in London: a resource for the Architecture and Design community.
material-lab.co.uk

Ingredients
Magazine designed and edited by Chris Lefteri, geared to informing and inspiring the work of designers.
moreingredients.com

100% Design 2012
The largest single-site design event in the UK—an exhibition at Earls Court featuring new materials.
100percentdesign.co.uk

WEBSITES CONTINUED

MIT Media Lab
MIT's research consortia for new technologies used in design.
media.mit.edu

Polymer and Composites Engineering Department (PaCE)
Research team at Imperial College London, seeks to bridge the gap between basic Chemistry, Materials Science, Engineering and Processing.
imperial.ac.uk/polymersandcompositesengineering

Next Nature
Speculative technologies that may become commercial within ten years, including nano technologies as well as recent materials innovations.
nextnature.net

Textile Futures Research Centre (TFRC)
Centre conducing textile related research across the University of the Arts, London—explores the role of textile industry in facilitating a sustainable future.
arts.ac.uk

Matweb
Online resource centre for materials for materials research—lists material property data.
matweb.com

Material Sense
Shared expertise and online collaboration focusing on cutting edge and new materials and products, and the exploration of the properties of materials.
materialsense.com

The Interactive Institute
Pioneering and experimental Swedish Design and Technology research institute that conducts world-class applied research and innovation.
tii.se

Surface Thinking
New materials blog powered by the Surface Design Show, looking at the inspiration and path for innovation of materials and ideas being developed.
surfacethinking1.blogspot.com

MATERIALS LIBRARIES

Advanced Materials Research Institute
College of Sciences
University of New Orleans
New Orleans
LA 70148
amri.uno.edu

A to Z of Materials
Suite 24, 90 Mona Vale Road
Warriewood
NSW 2101
Australia
azom.com

Centre for Materials Research, Queen Mary, University of London
Centre for Materials Research
Queen Mary, University of London
Mile End Road
London
E1 4NS
cmr.qmul.ac.uk

Environmental Design Research Society
Post Office Box 7146
Edmond, OK 73083 7146
USA
edra.org

Institute for Materials Research (IMR)
University of Leeds
Clarendon Road
Leeds
West Yorkshire
LS2 9JT
UK
engineering.leeds.ac.uk/imrl

Institute of Making, UCL
Malet Place
University College London
London
WC1E 6BT
materialslibrary.org.uk

Institute of Materials, Minerals and Mining
1 Carlton House Terrace
London
SW1Y 5DBUK
iom3.org

Library of 3-D Molecular Structures
nyu.edu

Lithuanian Materials Research Society
Sauletekio 10
LT-2040 Vilnius
Lithuania
ltmrs.lt

Material ConneXion Materials Library
60 Madison Avenue
2nd Floor
New York
New York 10010
Materialconnexion.com

Material Connexion, Cologne
Lichtstraße 43g
50825 Köln
Germany
de.materialconnexion.com

Material ConneXion, Milan
Via Carlo Poerio, 39
20126 Milano
Italy
it.materialconnexion.com

Materia, Netherlands
Pedro de Medinalaan 1b
1086 XK Amsterdam
The Netherlands
materia.nl

Materials Research Centre
ETH Zürich
Materials Research Center
HCIG 543
Wolfgang-Pauli-Sti. 10
CH – 8093 Zürich
Switzerland
mrc.ethz.ch

Materials Research Institute
N-317 Millennium Science Complex
Pollock Road
University Park
PA 16802
mri.psu.edu

Material Research Society
506 Keystone Drive
Warrendale, PA 15086-7573
USA
mrs.org
NanoHub
nanohub.org

The Rematerialise Project
School of 3D Design
Knights Park Campus
Kingston University
Kingston-upon-Thames
Surrey
KT1 2QJ
kingston.ac.uk

Royal Society of Chemistry—Chemsoc
Thomas Graham House
Science Park Milton Road
Cambridge
CB4 4WF
UK
chemsoc.org

Society for New Materials and Technologies in Slovakia (SNMTS)
C/o Slovak University of Technology
Department of Materials and Technologies
Pionierska 155k- 83102
Bratislava
Slovak Republic

Svenska Föreningen för Materialteknik
Dept of Materials Science & Engineering
Royal Institute of Technology KTH
SE-10044 Stockholm
Sweden

UCL Centre for Materials Research
University College London
Gower Street
London
WC1E 6BT
ucl.ac.uk/cmr

Material Lab
10 Great Titchfield Street
London
W1W 8BB
Material-lab.co.uk

AUTHOR BIOGRAPHIES

PHILIP HOWES

Philip Howes is a research scientist with a multidisciplinary work history and a keen interest in materials and making. After reading physics as an undergraduate, he undertook a PhD in nanotechnology at King's College London, which involved strong elements of physics, chemistry and biology. Since then Philip has been working with The Institute of Making on developing a sensoaesthetic theory of materials, a project which has involved using a combination of materials science and psychophysics to investigate people's interactions with the material world. A fundamental aim of this work has been to forge links between the materials science and materials arts communities by creating a commonality in their ways of thinking. Philip also has a strong interest in science writing and public engagement, and has been keen to communicate his work, and to help promote science as a whole, with various written articles and public events.

ZOE LAUGHLIN

Zoe Laughlin is a co-founder/director of the Institute of Making and the Materials Library project at University College London. She holds an MA from Central Saint Martin's College of Art and Design and obtained a PhD in Materials within the Division of Engineering, King's College London. Working at the interface of the science, art, craft and design of materials, her work ranges from formal experiments with matter, to materials consultancy and large-scale public exhibitions and events with partners including Tate Modern, the Hayward Gallery, the V&A, ICA, Wellcome Collection and Turner Contemporary. Her particular areas of interest are currently The Sound of Materials, The Taste of Materials, The Performativity of Matter and the relationship between materials and objects, with outputs ranging from theatrical demonstration lectures, peer reviewed papers, bespoke material-objects and features on both radio and television.

ACKNOWLEDGEMENTS

First and foremost, thanks to all the artists, designers and companies who were so forthcoming with images and information for the book, without which such a project would not have been possible.

In particular, thanks to James Carpenter Design Associates, Yemi Awosile, Jolan van der Wiel, Drummond Masterson, The Danish Design School, Diffus, Oskar Zieta, Doris Sung, Phillip Beesley and his team, Fraunhofer Institute, Markus Kayser, JM Architecture, Matthew Szosz, Susan Plum, Corning Inc, Marco Cevat, Swarovski Crystal Palace, Marc Newson, Yuya Ushida, Hovding, EADs, Christof Schmidt, Phil Cuttance, Oscar Wanless, Attua Aparicio, Oliver Poyntz, Ross Lovegrove, Jane ni Dhulchaointigh, Alyce Santoro, Beat Karrer, Mieke Meijer, Fabrican, Studio Koya, NEXT Architects, Jerszy Seymour Design Workshop, ShapeShift, Alexa Lixfeld, Concrete Canvas, Leeds University and Encos Ltd, Sachiko Kodama, LORD Corporation, Timm Herok, Micher'Traxler, Sarat Babu, Helen Storey, Michael Eden, James King, James Auger and Jimmy Loizeau.

Many, many thanks to Philip Howes and Zoe Laughlin for all their time, advice, assistance and for sharing their wealth of knowledge with us. Thanks also for our invaluable time spent with many of these materials in the Materials Library at the Institute of Making—a great place for anyone interested in materials. More information can be found at www.instituteofmaking.org.uk

At Black Dog Publishing, thanks to Leonardo Collina for his great and elegant design and Monica Oliveira for all her assistance with this, and to Phoebe Adler, Tom Howells, Emma Harwood, Arrate Hidalgo and Alice Lees for all their invaluable editorial assistance.

Black Dog Publishing Limited
10A Acton Street
London
WC1X 9NG

t. +44 (0)207 713 5097
f. +44 (0)207 713 8682
e. info@blackdogonline.com
www.blackdogonline.com

Designed by Leonardo Collina with the assistance of Mónica Oliveira at Black Dog Publishing.

Edited by Phoebe Stubbs.

Cover image
A sample of auxetic foam produced by additive layer manufacturing by BREAD Ltd's research project called Designing Hybrid Materials. The project is featured on page 214. Image courtesy BREAD Ltd.

British Library Cataloguing-in-Publication Data.
A CIP record for this book is available from the British Library.

ISBN 978 1 907317 73 6

Black Dog Publishing is an environmentally responsible company. Material Matters New Materials in Design is printed on FSC accredited paper.